Electronic Control Systems
Simulations and Experiments

ZAP Studio, LLC

Cover

Electronic Control Systems
Simulations and Experiments

by Sid Antoch

ZAP Studio, LLC · · · Philomath, Oregon

ISBN: 978-1-935422-13-6

Published by:

ZAP Studio, LLC
PO Box 1150
Philomath, OR 97370
www.zapstudio.com

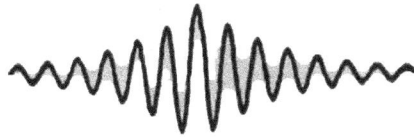

Contents

Resources

Parts
www.mouser.com
www.digikey.com
www.jameco.com
www.newark.com
www.alliedelec.com
www.allelectronics.com

Equipment
www.agilent.com
www.tek.com

USB Instruments

www.easysync-ltd.com
www.picotech.com

Simulation Software

Download free *LTspice* simulation software from
Linear Technology: www.linear.com

Introduction

This book may be used by any reader who wishes to learn by example and experiment. Simulation examples are presented which may be done using *LTspice*, a simulation program available as a free download from *Linear Technology*. Experiments provided may be performed using a solder-less breadboard, inexpensive parts, a small power supply, and a digital or USB oscilloscope. Some experiments also require a function generator.

All of the experiments demonstrate basic electronic control system principles. The experiments may be easily modified and may serve as the basis for other applications. Analysis suggestions are provided at the end of each experiment. This book may also be used as a supplement to a sophomore level electronic technology course with textbooks such as: *Modern Control Technology, by* Christopher Kilian, and *Industrial Electronics* by James A. Rehg and Glen J. Sartori.

The reader should be familiar with electronic devices such as transistors and operational amplifiers. A summary of solid state device properties and specifications is provided in chapter 1.

Although there are a few sections where a knowledge of calculus and Laplace transforms would be useful, these sections are not essential. Control system analysis will be based on a system's response to a variation of the system's parameters. System responses will be compared to simulation results and to theoretical expectations.

Electronic control system technology has a very broad range of application. This book focuses on the fundamentals which are applicable to many applications.

Sid Antoch

Chapter 1: Review of Control Devices

This chapter is intended to be used as a review and reference. It also presents the level of understanding required to do the experiments in this book. Device characteristics such as power dissipation and current handling capability are particularly important.

Electro-magnetic Relay

The relay is an electro-mechanical switch that is activated by a magnetic field. Relay switches have some advantages over transistor switches. Their contact resistance is very low and their contacts are isolated from the control circuit. However, the contacts are subject to arcing and wear. Also, relays operate much slower than transistors.

The contacts of the relay shown on the right are in the "de-activated" position. This is a single-pole double-throw relay. The common contact C is connected to contact 1. When the relay is activated by a current through the coil, contact C disconnects from contact 1 and connects to contact 2. Relays are available in a large range of power ratings and contact configurations.

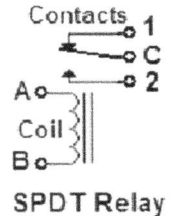

HAMLIN HE3621A1210 relay specifications:

Specification	Value
Coil Current	12mA
Coil Resistance	1000Ω
DC Coil Voltage	12V
Maximum Pick-Up Voltage	8VDC
Minimum Dropout Voltage	1VDC
Maximum DC Voltage	200V
Maximum Current Rating	0.5A
Maximum Power Rating	10W
Operate Time - max.	1mS
Release Time - max.	1mS

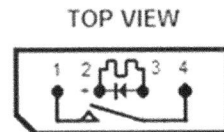

TOP VIEW

This relay has a diode across the coil to protect it against high transient reverse voltages. This relay's coil is rated at 12 volts and the specifications indicate that a minimum of 8 volts is required to operate it. When activated, the voltage across the coil must be reduced to less than 1 volt to deactivate it.

1

A relay is activated by the current in its coil which creates a magnetic field that operates the switch. This action takes time to complete and is specified as the "Operate Time – maximum" in the table above. The deactivate time is specified as the "Release Time – Maximum" in the table above. This timing is further investigated in experiment 1.

Bi-Polar Transistor

A bi-polar transistor can be used as a controlled voltage or current source as well as a switch. The basic parameters needed to understand and design transistor circuits are the transistor's current gain, β, and the base to emitter voltage, V_{BE}.

Figure A shows a voltage controlled voltage source. The transistor's output voltage, V_E, is controlled by its base voltage V_B.

Equations: $V_E = V_B - V_{BE} \approx V_B - 0.7$

$$I_E = \frac{V_E}{R_E} \qquad I_B = \frac{I_E}{\beta + 1}$$

Figure B shows a typical transistor switch circuit. When the transistor is off the current through R_C is zero and V_C is equal to V_{CC}. When the Transistor is on almost all of V_{CC} is dropped across R_C and V_C is close to zero. The actual value of V_C is equal to the transistor's saturation voltage, $V_{CE(SAT)}$.

Equations: $$I_C = \frac{V_{CC} - V_{CE(SAT)}}{R_C} \approx \frac{V_{CC}}{R_C}$$

$$I_B = \frac{V_{BB} - V_{BE}}{R_B} \approx \frac{V_{BB} - 0.7}{R_B}$$

Figure C is a voltage controlled current source.

Equations:
$$V_C = V_{CC} - I_C R_C$$

$$I_C = \frac{V_E}{R_E} = \frac{V_B - V_{BE}}{R_E} \approx \frac{V_B - 0.7}{R_E}$$

Note: V_{CE} must be greater than $V_{CE(SAT)}$.

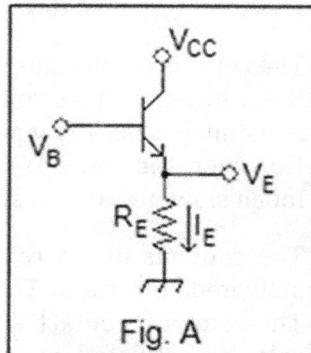

Fig. A

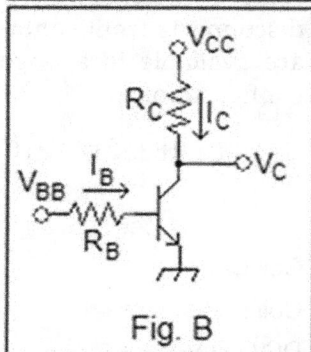

Fig. B

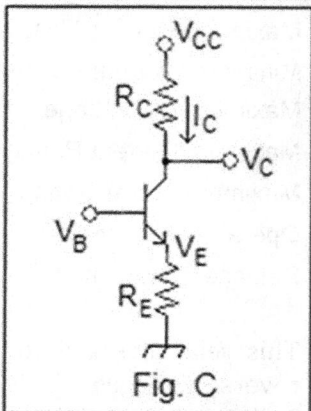

Fig. C

Power Mosfet

Enhancement mode power mosfets are used as switches in control systems. Refer to the graph and the diagram in figure D below. Voltage applied between the gate and source, V_{GS}, controls the drain current. The graph below shows the drain current versus drain-to-source voltage for various values of V_{GS} for the International Rectifier IRF510.

Note that when V_{GS} is 5.5 volts, a drain current of 0.5A results in a V_{DS} of 0.3V. This is in the saturation region, $V_{DS(SAT)}$, for the transistor.

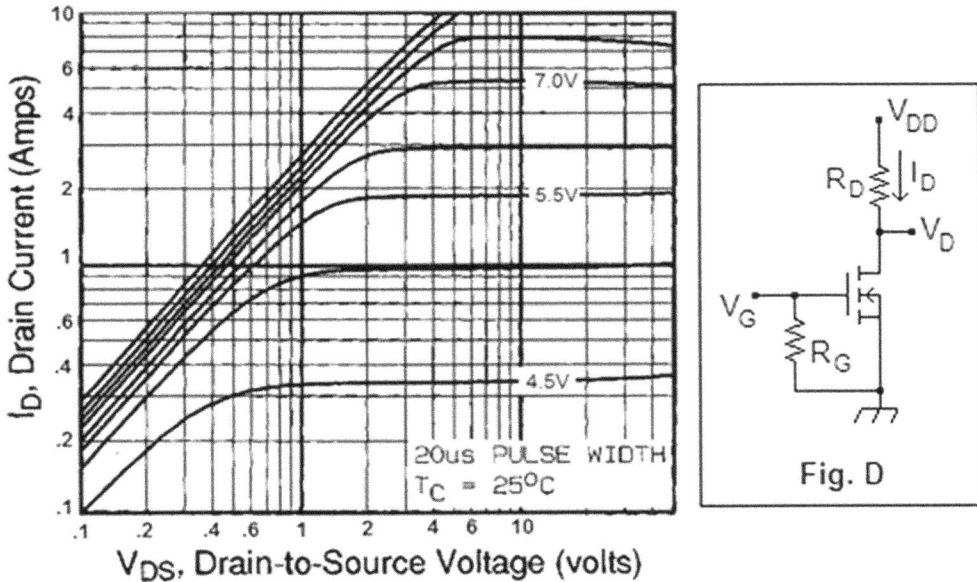

Fig. D

Equations for switch mode enhancement mosfet operation:

$$I_D = \frac{V_{DD} - V_{DS(SAT)}}{R_D} \approx \frac{V_{DD}}{R_D} \qquad V_D = V_{DD} - I_D R_D = V_{DS(SAT)} \approx 0$$

Note that the approximate value of $V_{DS(SAT)}$ can be obtained from the I_D versus V_E graph.

Unijunction Transistor

Applications of the unijunction transistor, UJT, include oscillator circuits, timing circuits, and trigger circuits. When the UJT's emitter voltage is below the UJT's specified peak voltage, Vp, the UJT's emitter resistance is very high.

3

However, when V_E reaches Vp, the emitter resistance decreases, resulting in an increasing emitter current and a decreasing emitter voltage as shown in the graph in figure F on the next page. This is called the "negative resistance" region because the emitter current increases while the emitter voltage decreases.

An RC circuit is connected to the UJT's emitter in the diagram in figure E on the right. The capacitor will charge until its voltage reaches Vp, at which point it will discharge, creating a current pulse through V1. Then V_E decreases and the capacitor begins charging again, repeating the cycle.

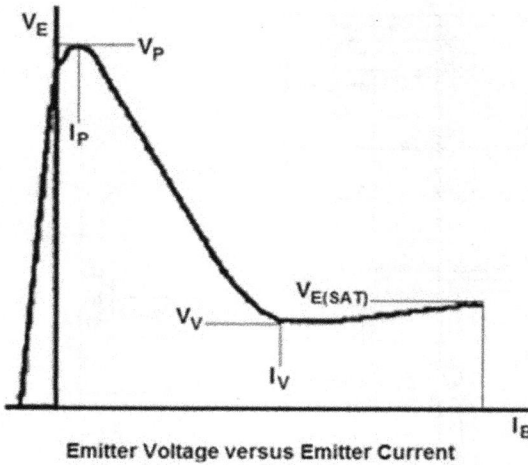

Emitter Voltage versus Emitter Current

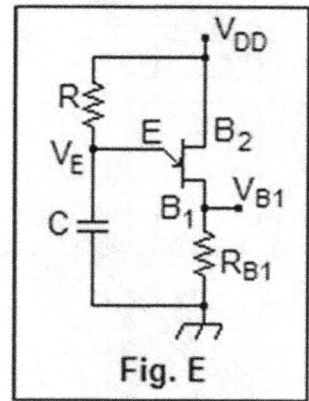

Fig. E

Thyristors: SCR and TRIAC

The SCR (Silicon Controlled Rectifier) is basically a diode that is turned on or enabled by a trigger voltage, V_{GT}, applied between the gate, G, and the terminal MT1. The trigger voltage is typically a short pulse of about 1 volt in magnitude. The SCR will continue to conduct until the SCR's anode current drops below its holding value, I_H.

The TRIAC is an AC version of the SCR. It is called a four-quadrant device because it can operate with either polarity of terminal voltage or gate voltage. Once triggered, the TRIAC will continue to conduct until its terminal current drops below the holding current, I_H.

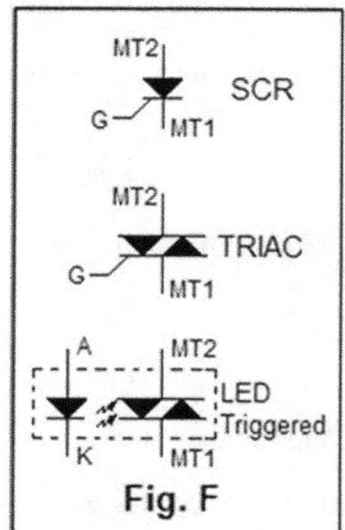

Fig. F

4

LED triggered TRIACs offer isolation between the trigger circuit and the controlled circuit. SCR's and TRIACS are available in a large range of current (mA to kA) and power (mW to kW) capability.

Operational Amplifier

All the op-amps in this section have a positive power supply voltage indicated as V_{CC} and a negative power supply voltage indicated as V_{EE}. The power supply connections are not shown except for U1 in figure G. The op-amp output saturation voltages are given as $V_{CC(SAT)}$ and $V_{EE(SAT)}$.

Comparator

U1 in figure G compares two analog input voltages. Its output voltage indicates which input voltage is greater.

Equations:

If Va > Vb, then Vo = $V_{EE(SAT)}$

If Va < Vb, thenVo = $V_{CC(SAT)}$

Inverting Amplifier

U2 in figure G amplifies the input voltage by the amplification factor, A_V.

Equation:

$$Vo = A_V V_{in} \qquad A_V = -\frac{R_f}{R_{in}}$$

Non-inverting Amplifier

U3 in figure G amplifies the input voltage by the amplification factor, A_V.

$$Vo = A_V V_{in} \qquad A_V = \left\{ 1 + \frac{R_f}{R_{in}} \right\}$$

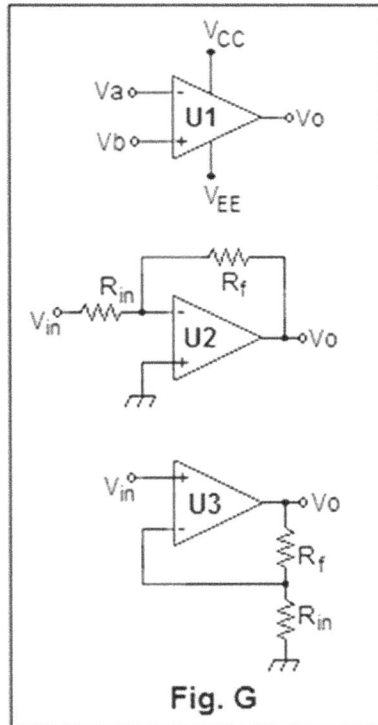

Fig. G

Summer

The output, Vo, of U1 in figure H is the inverted sum of the input voltages, V_1, V_2, and V_3 as given by the equations below:

$$Vo = -\frac{R_f}{R_1}V_1 - \frac{R_f}{R_2}V_2 - \frac{R_f}{R_3}V_3$$

$$Vo = -(V_1 + V_2 + V_3), \text{ if } R_1 = R_2 = R_3 = R_f$$

Integrator

The approximate output voltage, Vo, of U2 in figure H is given by the equations below. Note that $V_{(t=0)}$ is the output voltage of U2 at the start of the integration (at t = 0).

$$Vo \approx -\frac{1}{R_{in}C}\int_0^t V_{in}(\tau)\,d\tau + V_{(t=0)}, \ (R_f \gg R_{in})$$

$$Vo \approx -\frac{1}{R_{in}C}V_{in}\Delta t + V_{(t=0)}, \ (V_{in} \text{ a cons} \tan t)$$

Differentiators

The approximate output voltage, Vo, of U3 in figure H is given by the equation below.

$$Vo \approx -R_f C\frac{dv_{in}}{dt}, \ \text{ if } R_f \gg R_{in}$$

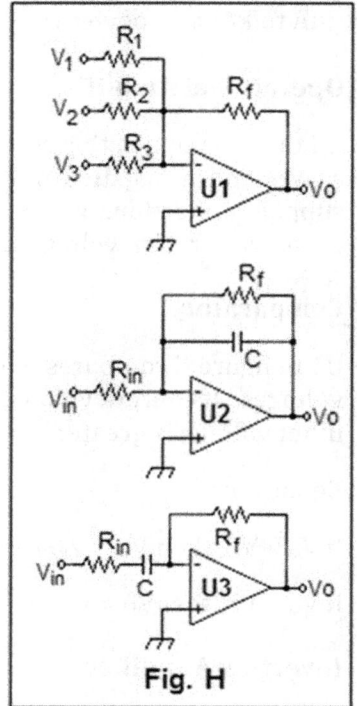

Fig. H

6

Chapter 2: Open Loop Experiments and Simulations

Experiments in this chapter involve basic components common to many electronic control systems: transistors, operational amplifiers, thyristors, and relays. These experiments may be used as the laboratory portion of a control systems course. They present a variety of applications which emphasize important control system concepts. In addition to the ability to use electronic laboratory equipment, knowledge of basic transistor and operational amplifier theory is assumed

Experiment 1: Lamp Blinker / Electromagnet Relay

This experiment uses an op-amp square wave oscillator and an electromagnetic relay as a lamp blinker. The blinking frequency is set by an RC time constant in the Schmitt trigger oscillator circuit. The operating characteristics of the relay will be measured and compared to the relay specifications. In particular, the relay's activation time, deactivation time, and contact bounce will be observed and measured.

Parts and Equipment Required

Op-Amp U1: LM358. Transistor Q1: 2N3904, Lamp: #2182 (14V, 80mA)
Relay: 12VDC SPDT Relay (600 to 1000 ohm winding),
Resistors: 4 each 10K, 1 each 470k, all ¼ Watt, 5%.
Capacitor: 1uF, 16V minimum, 5% or 10%, <u>non-polarized</u>.
Capacitor: 100uF, 16V minimum.
Oscilloscope, DMM, 12 Volt Power Supply

Procedure

1.   Before connecting the circuit, measure and record the resistance of the relay coil, R_W. Calculate and record the relay current, I_W, when 12 volts is applied to the coil.

     R_W _____ I_W _____

2.   Look up the specifications for the transistor you will use (2N3904 or similar). Is the transistor capable of conducting the relay current? How much power, P_D, do you expect the transistor to dissipate if the on voltage, V_{CE}, is 0.3 volts?

     P_D _____

3. Use the ohmmeter to verify the operation of the relay. Measure the contact resistance between the contacts when relay is un-activated and when activated (12 VDC applied).

Relay	Common to NO, Ω	Common to NC, Ω
Un-activated		
Activated		

4. Connect the circuit below and turn on the power.

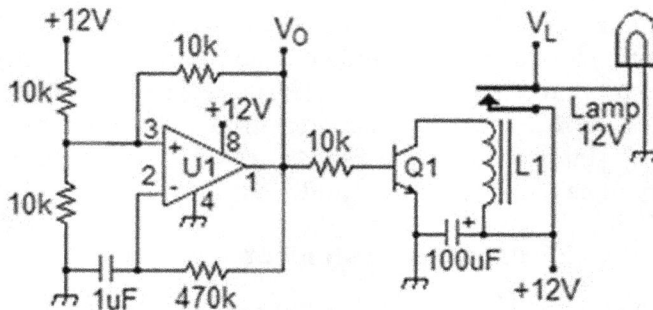

5. The lamp should be flashing on and off about once per second. Connect channel 1 of the oscilloscope to V_O and channel 2 to V_L.

6. Set the trigger to channel 1 and positive slope. Set channel 1 to 5V/DIV and channel 2 to 2V/DIV with DC input coupling. Set the zero reference for both to the bottom of the screen.

7. Set the time base to see at least one cycle of the channel 1 waveform (V_O). Carefully observe the timing relationship between the waveforms, V_O and V_L.

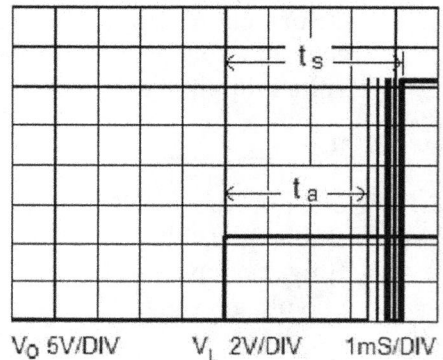

8. Adjust the trigger and the time per division to make an accurate measurement. The graph on the right shows V_O rising from 0 to 10 volts at the center of the graph.

V_O 5V/DIV V_L 2V/DIV   1mS/DIV

8

The relay contacts close at about 8.5 milliseconds. You can observe about 1 millisecond of "contact bounce" on channel 2, which is due to the vibration of the relay contacts.

9. Measure and record the length of time it takes to activate the relay contacts, t_a, and the length of time it takes for the contacts to settle, t_s, (after the contacts stop bouncing). The graph on the right shows an activation time of about 3.5ms and a settling time of about 4.2ms. The settling time here refers to the total time between the activating signal and the last contact bounce (4.2 milliseconds in the graph above).

t_a _____ t_s _____

Analysis:

1. Obtain the specifications for the relay used in this experiment and compare your results to the specifications (relay current and settling time).

2. Determine the expected frequency of the Schmitt trigger oscillator. Compare your results.

3. Calculate the duty cycle of the lamp and the efficiency of the light blinker circuit.

$$\text{Duty Cycle} = \frac{\text{On time}}{\text{On time} + \text{Off time}} \qquad P_{Lamp} = P_{Lamp(on)} (\text{DutyCycle})$$

$$\%\text{Eff} = \frac{P_{Lamp}}{P_{Lamp} + P_{Circuit}} \cdot 100\%$$

Experiment 2: Pulse Width Modulation and Power Control

Power delivered to a load, such as a lamp, heater, or dc motor, can be most efficiently controlled using pulse width modulation (PWM). A power transistor connected in series with the load is repetitively turned on and off. When the transistor is on, almost all of the power supply voltage is supplied to the load. When the transistor is off, no power is supplied to the load.

Enhancement mode power mosfets are often used to supply power to a load using PWM. The average power delivered to the load is controlled by the "duty cycle" of the transistor, which is the ratio of its "on" time to the time of one cycle. $I_{DS(ON)}$ is the transistor's drain to source current when it is on, and $V_{DS(ON)}$ is the transistor's drain to source voltage when it is on. Given that DC is the duty cycle, the average power, P_L, supplied to a load, and percent efficiency, %Eff, are given by:

$$P_L = I_{DS(ON)}(V_{DD} - V_{DS(ON)}) \cdot DC \qquad \%Eff = \frac{P_L}{(P_L + P_Q)}100\% = \left[1 - \frac{V_{DS(ON)}}{V_{DD}}\right]100\%$$

$$\text{where:} \quad I_{DS(ON)} = \frac{(V_{DD} - V_{DS(ON)})}{R_L} \qquad P_Q = I_{DS(ON)} \cdot V_{DS(ON)} \cdot DC$$

Refer to the diagram on the right. The transistor is turned on and off by the pulse width modulator, which consists of an analog comparator and a triangle wave oscillator. A triangle wave is applied to one input of the comparator and a control voltage to the other.

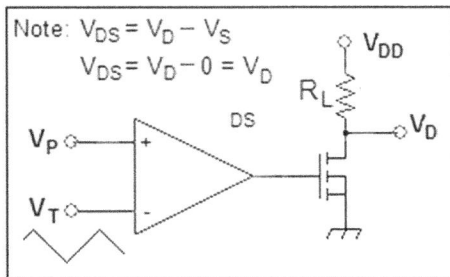

When V_P is greater than V_T, the output of the comparator is high and the transistor is on. When V_P is less than V_T, the output of the comparator is low and the transistor is off. The duty cycle and average power delivered to the load can be varied by varying V_T.

The triangle wave source for this experiment is a simple "Schmitt Trigger" type op-amp relaxation oscillator, shown on the right. It produces a 4V p-p. 700 hertz, approximation of a triangle wave.

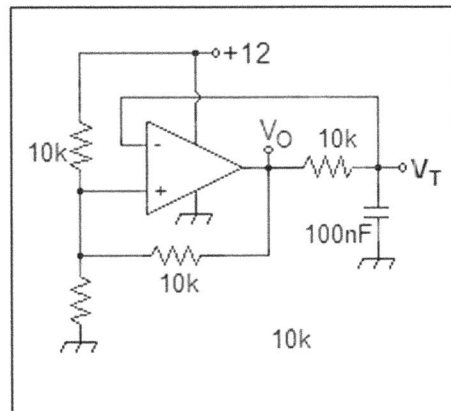

When V_O is high, about 10.5 volts, the capacitor charges. When V_O is low, near zero volts, the capacitor discharges. The resistor values shown produce a trigger voltage at the non-inverting input of 8 volts when V_O is high and 4 volts when V_O is low. The result is that the capacitor charges exponentially to 8 volts and discharges exponentially to 4 volts.

The resulting waveform approximates a triangle wave because the charge and discharge times are less than one RC time constant.

This exercise will demonstrate the ability of the mosfet to vary the amount of power supplied to the load using PWM (Pulse Width Modulation). The circuit is basically a "light dimmer", but the same principle can be used to control power supplied to motors and to other devices

Parts and Equipment Required

Power supply: +12VDC. Oscilloscope. DMM. Lamp: #2182 (14V, 80mA)
Op-amp: LM358. Transistor: IRF511 MOSFET.
Resistors: 8 – 10k, all ¼ watt, 5%.   Resistor: 100Ω, 5 watt, 5%.
Potentiometer: 10k one-turn trimmer.
Capacitors: 100nF, 50V, 5%.   100µF, 25V, Electrolytic.

Procedure

1.   Connect the circuit below.

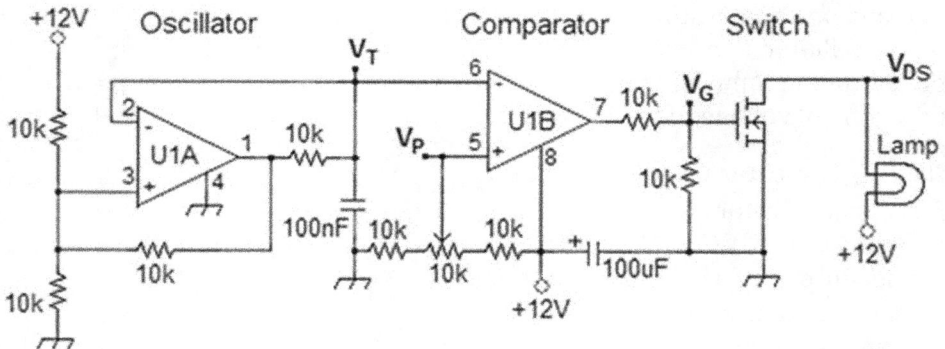

Refer to the picture of the mosfet on the right. Note that the mounting tab on the mosfet is also its drain.

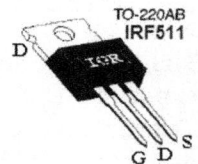

2.   Verify that the circuit is working by adjusting the 10K pot. The pot should vary the brightness of the lamp between zero and maximum.

3. Connect channel 1 of the oscilloscope to V_T and channel 2 to V_G (mosfet gate). Connect the DMM to measure the DC voltage V_P on pin 5 of U1. Adjust the 10K pot to set V_P to 6.0 volts.

4. Set channel 1 and channel 2 of the oscilloscope to 1V per division and DC input coupling.

   Use vertical centering controls to set the 0 reference for both traces to bottom of the screen. Set the trigger to channel 2 and positive slope. Set the time base for 500 microsecond per division. Refer to the display on the right.

   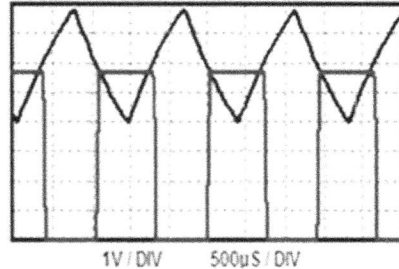

   1V / DIV   500μS / DIV

5. The percent of time that TP1 is high is called the "duty cycle" of the square wave. Vary the voltage V_P on pin 5 of U1 and note how the waveform's duty cycle changes. You should be able to adjust it from 0% to about 100%. Note how the brightness of the lamp varies with the duty cycle.

   PWM (Pulse Width Modulation) is a very efficient way of controlling power supplied to devices such as motors, heaters, and lamps.

   $$\% Efficiency = \frac{Power\ Output\ by\ Control\ Circuit}{Power\ Input\ to\ Control\ Circuit} \cdot 100\%.$$

6. Replace the lamp with a 100Ω, 5 watt resistor. Set the voltage V_P on pin 5 of U1 to 6.0 volts with the 10K pot. This should result in a 50% duty cycle.

7. Connect Channel 1 of the oscilloscope to V_{DS}. Leave channel 2 connected to V_G and leave the trigger on channel 2. Set channel 1 to 50mV per division. The positive peaks will be off the screen. The objective here is to measure the voltage across the transistor when it is on ($V_{DS(ON)}$). Adjust the both channel 1 so that you can accurately measure the value of $V_{DS(ON)}$.

8. Repeat measurements for duty cycles of 25% and 75% and record your results below.

   $V_{DS(ON)}$ 25%DC_____ $V_{DS(ON)}$ 50%DC_____

   $V_{DS(ON)}$ 75%DC_____

13

Analysis:

1. Transfer the results of step 8 into a spreadsheet. Use the table layout shown below. $I_{DS(ON)}$ is the transistor current when the transistor is on. P_Q is the average power dissipation of the transistor. P_L is the average power dissipation of the 100 ohm resistor. DC is the duty cycle expressed as a fraction.

DC	$V_{DS(ON)}$ Volts	$I_{DS(ON)}$ Amps	P_Q Watts	P_L Watts	R_L Ohms	Efficiency Percent
.25						
.50						
.75						

2. An easier way to vary the power delivered to a load would be to connect a variable resistor (rheostat) between the battery and the load.

    Compare the efficiency of the PWM circuit in delivering power to a 100 ohm resistor operating at 25% and 75% duty cycle to the rheostat circuit shown on the right.

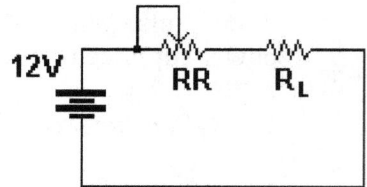

    (a) Calculate the current, I, required to deliver the same power to the 100 ohm resistor in the DC circuit as was delivered to the lamp in the PWM circuit for a 25% and 75% duty cycle.

    (b) Calculate the value of the resistance RR required to obtain the currents calculated above for each duty cycle..

    Hint: Use $P_L = I^2R_L$ to calculate I in the DC circuit. Power input to the circuit is 12V times the circuit current, I. ($I = 12/(RR + R_L)$)

3. Use the results above to compare the efficiency of the rheostat circuit above to the 25% duty cycle and 75% duty cycle PWM circuit.

Pulse Width Modulation and Power Control Simulation

The LTspice diagram is shown on the right. The LT1006 op-amp is equivalent to the LM358.

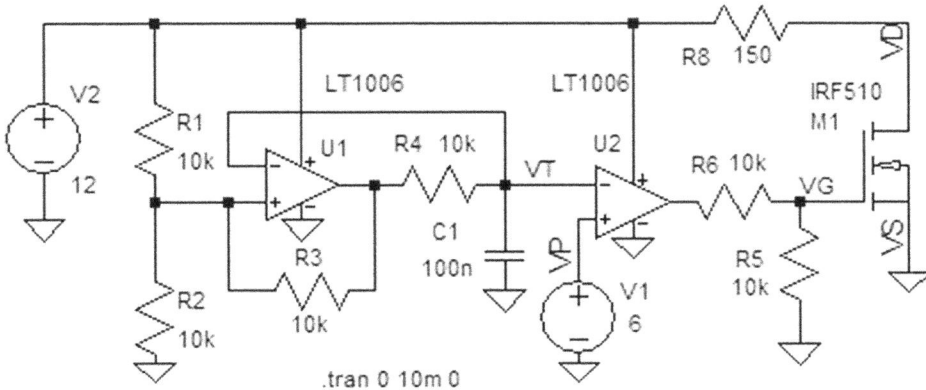

The IRF510 was obtained by first selecting "mosfet" and placing it into the schematic. Right click on the mosfet, select "Pick New MOSFET" and scroll down until you find the IRF510.

A 150 ohm resistor, R8, represents a 2182 lamp. Simulation was set to "Transient" with a stop time of 10 milliseconds. The graph below on the left shows the triangle wave output of the oscillator and the waveform at the gate of the transistor with V_P set to 6.0 volts.

The graph below on the right shows the waveform on the drain of the transistor. When the transistor is on, about 53mV appears on the transistor's drain. This is the transistor's saturation voltage. V_{DS}. When the transistor is off, the drain voltage is 12 volts and is off the graph.

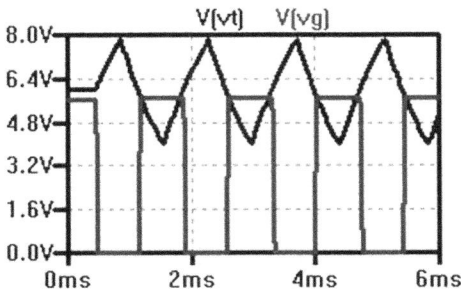

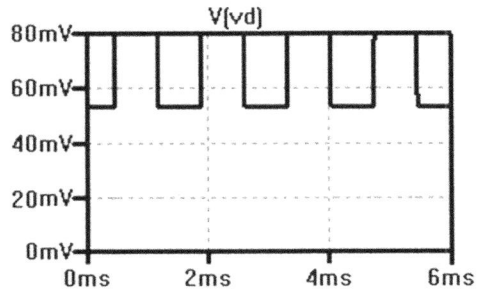

Experiment 3: AC Power Control / Triac

Triacs are devices that can control AC power. They are commonly used in home light dimmers in addition to controlling power to devices such as ovens and motors. Once triggered on, they will continue to conduct until the current decreases to less than the devices "holding current". They are ideal for AC power control because they can be triggered on at any phase angle of the input sinusoid and they automatically turn off at the sinusoid's zero crossings.

This lab exercise uses a "unijunction transistor" to trigger the triac at an adjustable phase angle. An optically triggered triac is used in this exercise. The control circuit used has two separate grounds that must be kept isolated from each other. Be sure to study the circuit below carefully before doing this experiment.

Parts and Equipment Required

AC Source, 12 VRMS, 200mA (Wall Transformer rated 200mA to 600mA).
Multimeter and Oscilloscope. Lamp: 14 Volt, 80mA (2182, 382, 386, or 756).
Opto-Triac: MOC3010 or NTE3047 or ECG3047.
Zener Diode: 5 to 6 Volt, ½ Watt. Unijunction Transistor: 2N4871.
4 Diodes: 1N4002 or equivalent. Capacitor: 0.1µF.
Resistors: 330, 10K, 2-1K, ¼ Watt, 5%. Pot: 100K, 1-turn, trimmer type.

Procedure

1. Connect the circuit below but do not apply power. Note that this circuit has two separate grounds that <u>must not</u> be connected together. Double check your connected circuit. If possible, have someone else (lab partner) check the circuit. "JPR" is a removable "jumper" wire.

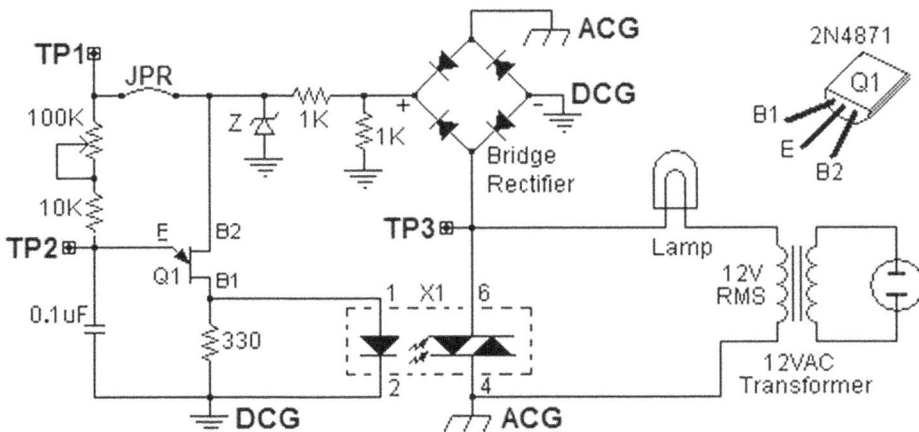

2. Apply power. Check that you can adjust the lamp brightness with the 100K pot. Set the pot for about one half of the maximum brightness.

3. The next set of measurements determine the timing relationship of waveforms at TP1, TP2, and TP3. Note that the circuit an AC ground (ACG) and a DC ground (DCG). <u>Do not</u> connect the oscilloscope to ACG and DCG at the same time.

4. Connect the oscilloscope ground leads to the DC ground, DCG. Connect channel 1 to TP1, channel 2 to TP2, and trigger on channel 1 on negative edge.

5. Adjust the time base so you can observe two to three cycles, similar to the display on the right. Measure and record the maximum value of the voltage V_Z at TP1.

$V_{z(max)}$ _____

TP2 Waveform is the capacitor voltage. The capacitor charges to about 4 volts and then rapidly discharges. The discharge of the capacitor triggers the triac. When the triac conducts the voltage to the bridge rectifier drops, causing TP1 voltage to drop.

6. Disconnect both channels of the oscilloscope from TP1, TP2, and DCG.

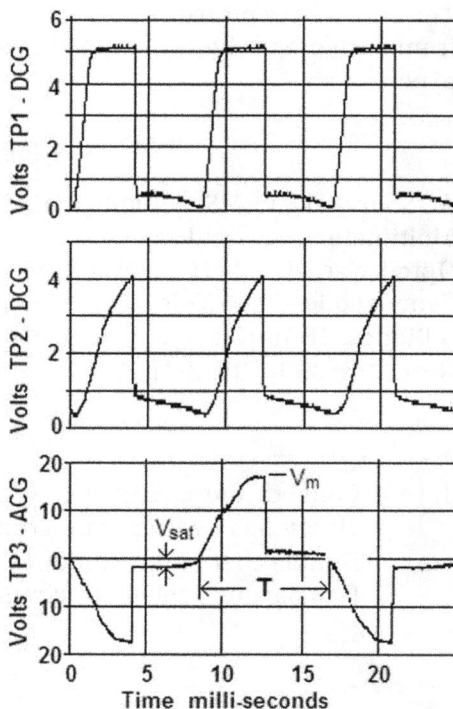

7. Connect channel 1 to TP3 and to the AC ground (ACG). Adjust the potentiometer so that the negative edge of the positive peak occurs at its maximum value, V_m, as shown on the right.

This results in a firing angle of about 90 degrees and a firing time of about 4.2mS.

Measure and record the peak to peak amplitude of the waveform.

V_m _____ Volts peak to peak.

18

8.   Measure and record the average peak to peak value of the saturation voltage, V_{sat}, at TP3. Increase the volts per division on the oscilloscope tp make an accurate measurement.

V_{sat} _____ Volts peak to peak.

9.   Disconnect channel 1 from TP3 and the AC ground. Connect channel 1 to TP2 and the DC ground. Measure and record the maximum (peak) capacitor voltage.

V_P _____ Volts peak.

10.   Disconnect the power, remove jumper wire, JPR, and measure and record the resistance between TP1 and TP2. Do not disturb the potentiometer setting. This measurement gives you the value of R_t for the RC time constant.

R_t _____

Analysis:

1.   Step 9 of the procedure determines the firing voltage, V_P of the Unijunction transistor. Step 10 determines the "charging" resistance R_t.

Use the measured values of V_P, R_t and the value of the capacitor to determine the approximate firing time, t_P.

$$V_P = V_Z (1 - e^{-\alpha t_P}), \text{ where } \alpha = \frac{1}{RC} .$$

2.   Compare the result to your measured t_P, which is about 4.17 mS as this is the time for one quarter cycle (90 degrees) of a 60 hertz sinusoid.

Calculate the percent accuracy of your measured result compared to your calculated result. Discuss the sources of error in your measurements and calculations.

3.   Calculate the efficiency of the triac using your measurements from the TP3 waveform. Consider that the triac conducts for 90 degrees of the input sinusoid for each half cycle of   the input sinusoid. Verify the equations below.   Note that the current was not measured, but it will divide out in the efficiency equation.

$$\% Eff = \frac{P_{Lamp}}{P_{in}} = \left[1 - \frac{2\sqrt{2}\,V_{SAT}}{V_{in(pp)}} \right] 100\%$$

Note: Expressions for input power, lamp power, and TRIAC power below:

$$P_{in} = \frac{V_{m(pp)} I_{RMS}}{4\sqrt{2}} \qquad P_{Lamp} = P_{in} - P_{Triac} \qquad P_{Triac} = \frac{V_{sat(pp)} I_{RNMS}}{2}$$

4.   Discuss the effect of increasing the transformer voltage on the efficiency of the triac. You   may refer to triac data sheets or on the internet.   The parameter that has the greatest effect on efficiency is the "Peak On State Voltage".

NTE3047 Electrical Characteristics: (TA = +25°C)

Parameter	Symbol	Test Conditions	Min	Typ	Max	Unit
Input LED						
Forward Voltage	V_F	$I_F = 10mA$	-	1.15	1.50	V
Reverse Leakage Current	I_R	$V_R = 3V$	-	0.05	100	μAA
Output Detector ($I_F = 0$ unless otherwise specified)						
Peak Blocking Current, Either Direction	I_{DRM}	Rated V_{DRM}	-	10	100	nA
Peak On-State Voltage, Either Direction	V_{TM}	$I_{TM} = 100mA$ Peak	-	1.8	3.0	V
Critical Rate-of-Rise of Off-State Voltage	dv/dt		-	10	-	V / μs
Coupled Characteristics						
LED Trigger Current, Output Latch Current	I_{FT}	Main Terminal Voltage = 3V,	-	8	15	mA
Holding Current	I_H		-	100	-	μA

Experiment 4: Switch Mode Voltage Boost Converter

Switch mode converters have become a very common voltage source. Many integrated circuits are available for switch mode power supplies which also include voltage regulation and current limiting. This exercise demonstrates the basic concepts of a switch mode converter without voltage regulation or current limiting.

The diagrams below show a voltage source, **Vi**, connected to an inductor in series with a transistor, Q.

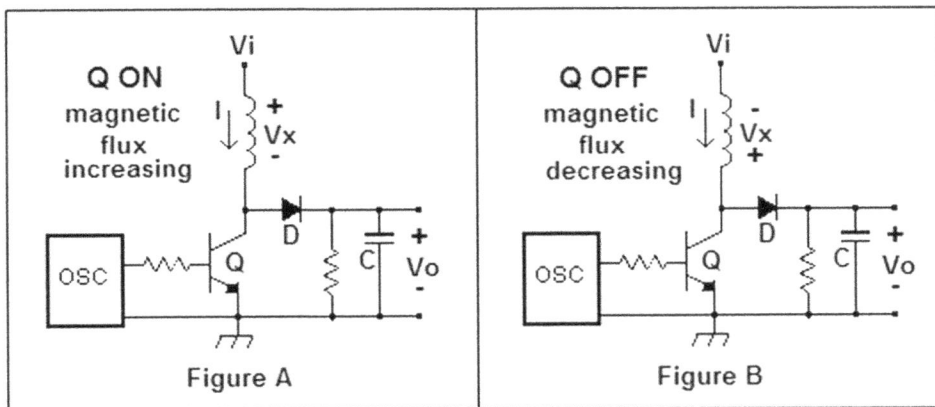

Figure A   Figure B

Figure A shows that when Q is on, the current, **I**, in the inductor increases as does the magnetic field. During this time no current flows through the diode, D. The output voltage is that of the capacitor, C.

Figure B shows that when Q is off, the current, **I**, in the inductor decreases as does the magnetic field. The collapsing magnetic field induces the voltage, **Vx**, across the inductor, as shown in the diagram. **Vx** adds to the supply voltage, **Vi**, and current flows through the diode, D, charging the capacitor to the voltage **Vo**. **Vo** = **Vi** + **Vx** - 0.7 volts.

The output voltage, **Vo**, can be controlled by varying the duty cycle of the switching transistor, Q. Special purpose ICs are available for switch mode power supplies which include a pulse-width modulated oscillator, voltage regulator, and current limiter.

This lab experiment uses an op-amp as a control circuit that varies the switching duty cycle and the output voltage of the converter.

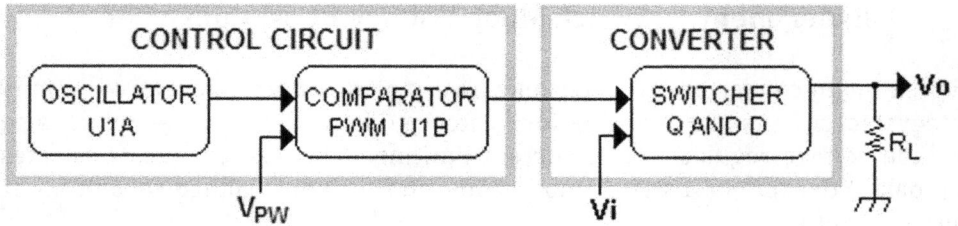

In the block diagram above, the oscillator generates a triangle wave for the pulse width modulator. The voltage V_{PW} controls the duty cycle of the square wave generated by the PWM circuit. The PWM square wave is applied to the converter. The DC input voltage, **Vi**, is converted to the DC output voltage, **Vo**.

Equipment and Parts Required

Power Supplies, 6 volt and 12 Volt, DMM, and Oscilloscope.
Diode: 1N4001 or equiv. 10k trim pot.
Resistors: 0.1, six 10K, 100K, all ¼ watt, two 100 ohm, 5 Watt, 5%.
IC: LM358, Capacitors: 10 µF, two 10nF, two 100µF, (25 to 50 volt).
Inductor: 1mH, 1.3 amp min., 0.4 ohm max. (J. W. Miller 2124-H-RC).

Procedure

1. Build the control circuit below. Connect a DMM to measure the voltage at **v2**. Set **v2** voltage (V_{PW}) with the potentiometer to exactly 6.0 volts.

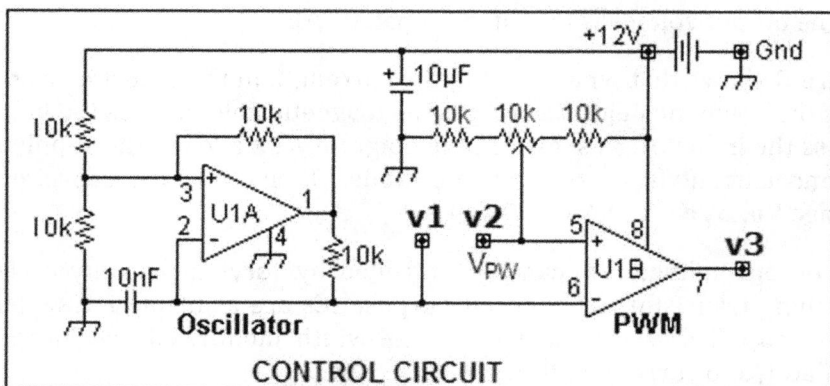

2. Connect oscilloscope channel 1 to **v3** and channel 2 to **v1**. Set both channels to 2 volts per division and time base to 50µS per division. Trigger on channel 1.

Observe an approximate square wave on channel 1 and triangle wave on channel 2 as shown on the right. You can see "slew-rate limiting" and a phase shift on v3 by the LM358 op-amp. A faster op-amp could have been used.

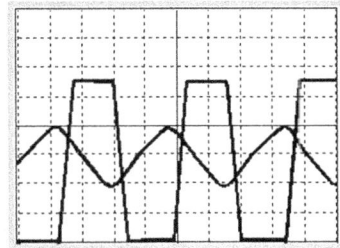

Measure and record the period of the waveforms and the maximum and minimum values of the triangle wave.

Period: _____   V$_{max}$: _____   V$_{min}$:_____

3. Set **v2** voltage (V$_{PW}$) with the potentiometer to exactly 6.0 volts. Measure and record duty cycle of the square wave (use the pulse width at 50% amplitude).

    Duty cycle (6V): _____

4. Set **v2** voltage (V$_{PW}$) with the potentiometer to exactly 5.0 volts. Measure and record duty cycle of the square wave. Set **v2** voltage with the potentiometer to exactly 7.0 volts.   Measure and record the duty cycle of the square wave.

    Duty cycle (5V): _____   Duty cycle (7V): _____

5. Connect the circuit below. R$_L$ = 100. The inductor should be close to the transistor. Do not apply 6 volt power to the converter yet.

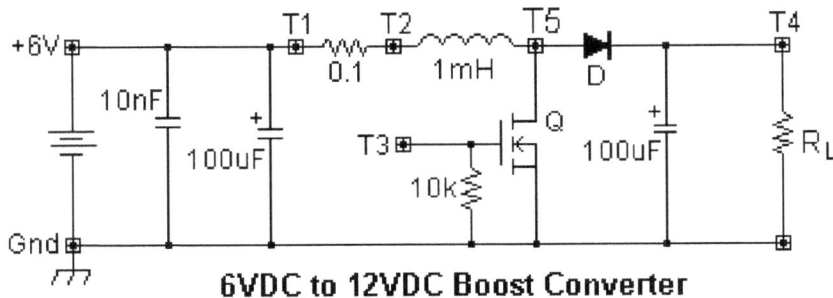

6VDC to 12VDC Boost Converter

It is important that the ground lead of the converter circuit (Transistor's source terminal) connects directly to the 6 volt power supply. Do not use the same ground lead as the 12 volt supply. Refer to the diagram on the right.

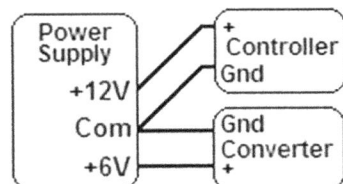

6.   Connect **v3** of the control circuit to T3 of the converter circuit. Connect oscilloscope   channel to T5 and channel 2 to T4 to monitor the converter output voltage. Trigger on channel 1.

7.   A more accurate value for the duty cycle can be obtained by measuring the actual time that the transistor is on.   When the transistor is on, the voltage at T5 will be close to zero and equal to the transistor's saturation voltage.

   Set v2 to 5.0 volts. Turn on the 6 volt power supply.   Measure and record the time T5 is low and the period.   Measure and record the transistor's saturation voltage, **V$_{SAT}$**.   Measure and record **Vo**.   Enter measurements into the table below. Turn off the 6 volt power supply.

v2 (V$_{PW}$)	T5 Low µSec.	T5 Period µSec.	Duty Cycle %	**V$_{SAT}$** Volts	**Vo** 100Ω Volts	**Vo** 50Ω Volts
5.0 volts						
6.0 volts						

8.   Connect an extra 100 ohm resistor in parallel with the already connected 100 ohm resistor so that the load resistance is 50 ohms. Turn on the 6 volt power supply.   Measure and record the output voltage for **v2** with the DMM as quickly as possible. Remove the extra 100 ohm resistor and turn off the 6 volt power supply when done.

   Set **v2** to 6 volts and repeat steps 7 and 8.

9.   Connect the extra 100 resistor so that the load resistance is 50 ohms. Connect oscilloscope channel 1 to T1 and channel 2 to T2. Set oscilloscope channel 1 and channel 2 to 20mV/Div and AC coupling. Set the oscilloscope to differential mode (channel 1 minus channel 2).   Connect the DMM in series with the 6 volt power supply to measure the supply current. Meter must be on the 2 amp scale.

10.   Turn on the 6 volt power supply. Record the power supply current, Is, and the peak to peak amplitude of the triangle wave across the 0.1 ohm resistor.

   Is: _____   V$_{T\ p\text{-}p}$: _____

Turn off the supplies. Wait for the 100 ohm resistors to cool before removing them.

Analysis

Analysis of the switch mode converter is based on the relationship between voltage and current of an inductor:

$$v=L\frac{di}{dt} \implies di=\frac{v}{L}dt \implies \Delta i=\frac{v}{L}\Delta t$$

The graphs on the right show the inductor voltage and current for a 50% duty cycle. Period of the 5KHz waveforms is 200μsec. Q is on for 100μsec and off for 100μsec.

$$i_{P-P}=\frac{6volts}{1000\mu H}100\mu\sec=0.6\,Amps$$

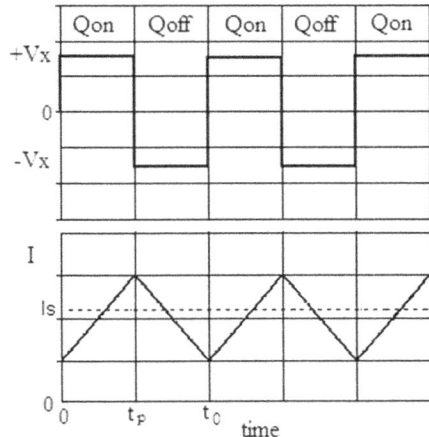

The current does not change direction. Vx is positive when the current is increasing and negative when the current is decreasing.

1. Calculate the frequency of the switching waveform, f_{sw}, from procedure step 2. Determine the values of the voltage at v2 which result in a 0% and a 100% duty cycle.

   f_{sw}: _____   **v2** 0%: _____   **v2** 100%: _____

2. Use the data in the table of procedure step 7 to calculate the theoretical peak to peak value of the inductor current, **i_{P-P}**, when v2 is 6 volts and R_L is 50 ohms.

   i_{P-P}: _____ (theoretical)

3. Calculate the measured value of the peak inductor current, i_{P-P}, from the measurements of procedure step 9.

   i_{P-P}: _____ (measured)

4. Calculate the efficiency of the converter with a 50 ohm load and **v2** = 6 volts. Use the value of the supply current, Is, to calculate the input power and the voltage across the load resistance to calculate the output power.

5.   Calculate the transistor's power dissipation when **v2** is 6 volts and R_L is 50 ohms.

Simulation of Switch Mode Voltage Boost Converter

LTspice is free from Linear Technology: http://www.linear.com. Below is the LTspice schematic. U1 is a high speed dual comparator, LT1720 from the comparator library. If you did the experiment, use the experiment part values for the simulation.

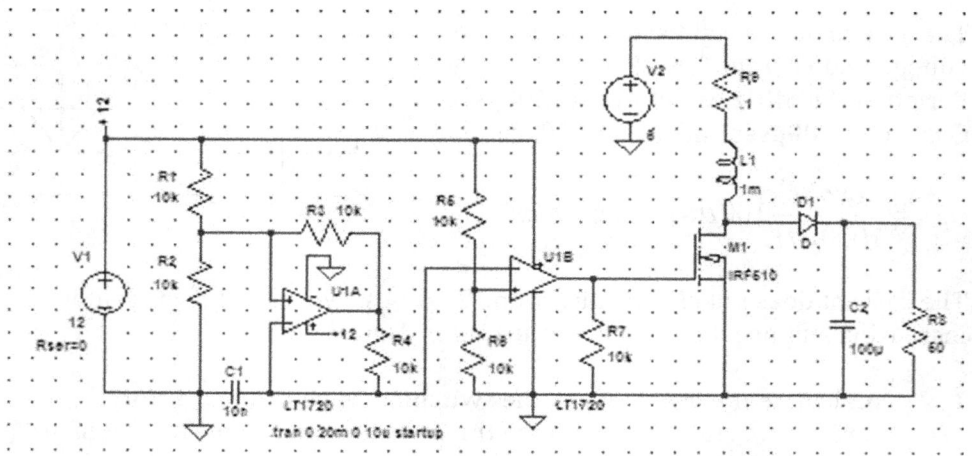

Simulation Results

The simulation was run for 15 milliseconds to allow time for the output voltage to settle. The graph on the next page shows an output voltage of about 11 volts and peak supply current of about 0.6 amps. The average supply current is about 0.44 amps with a peak to peak fluctuation of about 0.41 amps. The efficiency of the converter is:

$$\%Eff = \frac{Pout}{Pin} \cdot 100\% = \frac{11^2/50}{6(0.44)} \cdot 100\% = 92\%$$

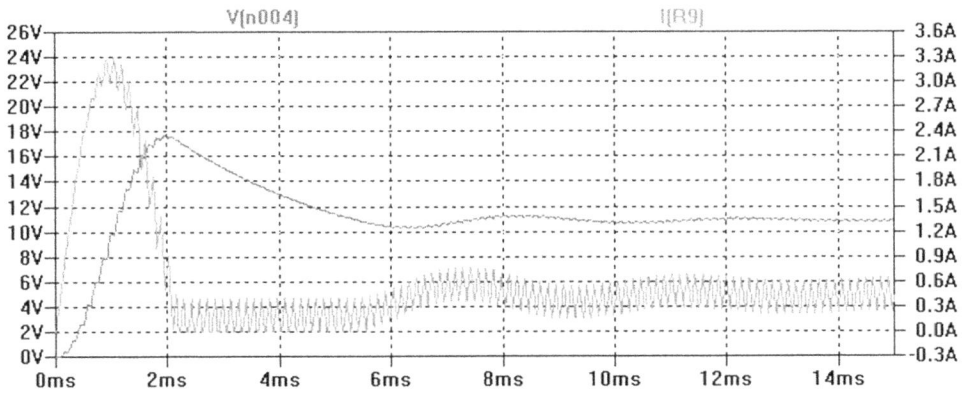

The graph above also shows the startup response of the converter. Note that it takes over 14mS for the converter to stabilize and reach steady state. It shows that the steady state output voltage is close to 11 volts and current is about 0.45 amps (average value of the triangle wave).

Chapter 3: Closed Loop

Control systems may be divided into two basic categories: open loop and closed loop. The block diagram below shows the basic components of all control systems. The control system controls a process, which may be the speed of a fan, the temperature of an oven, or the position of a camera, for example. The input to the controller is typically called the "set point". The set point is the desired output. The function of the controller is to achieve the desired output, the set point. The sensor monitors the actual output of the control system. This may be a tachometer or thermometer, for example.

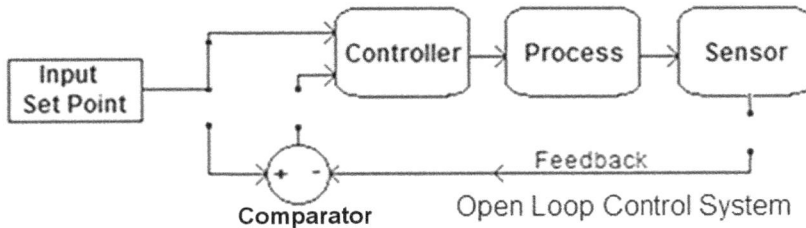

Open Loop Control System

A block diagram of an open loop system is shown above. An open loop system can be changed to a closed loop system by adding feedback. An automobile with cruise control is an example of a control system which may be either open loop or closed loop. With the cruise control off the driver controls the speed with the accelerator pedal. By observing the speedometer the driver is able to achieve the desired speed. If the driver is considered to be part of the loop the overall system with the driver is closed loop. However, to the driver the system is open loop.

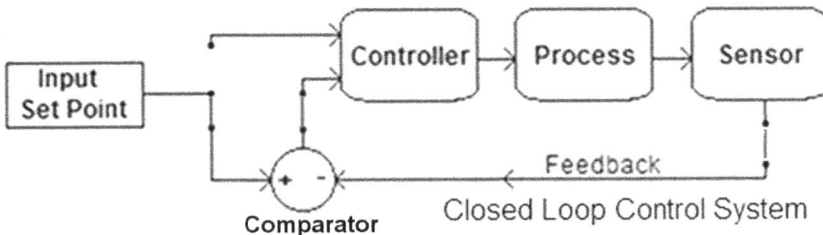

Closed Loop Control System

When the cruise control is enabled the speed is maintained automatically. The driver only needs to input the desired speed. The diagram above shows the added closed loop component. The comparator compares the output of the sensor (actual speed) to the input set point (desired speed). The output of the comparator causes the controller to maintain the set point speed. Closed loop systems will be discussed in more detail in the next chapter.

Basic Feedback Control Systems

Feedback control systems are common to mechanical, electrical, and biological systems. Examples include power supply voltage regulation, motor speed regulation, temperature regulation, pressure regulation, and measuring and positioning systems.

Feedback is used to compare the actual output of a control system to the desired output. In control system terminology, the actual output is usually referred to as the "process variable" and the desired output is the "set point". The difference between the process variable and the set point called the "error". The job of the control system is to reduce the error and to produce the desired output.

A simplified block diagram of a control system is shown on the right. **Vsp** is the set point, **Vpv** is the process variable, **Ve** is the error, and **H$_f$** represents the transfer function of the control system and control process.

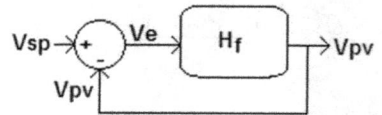

$$H_f = \frac{Vpv}{Ve} \qquad Ve = Vsp - Vpv$$

Feedback control systems vary in complexity, but generally fall into four categories: on-off, proportional, proportional plus integral, and proportional plus integral plus derivative. The intent here is to introduce the most basic of each type, including design principles and characteristics.

A control system is composed of a variety of components specific to its application. Each component contributes to its transfer function, **H$_f$**. The components typical to a control system are shown in the block diagram below.

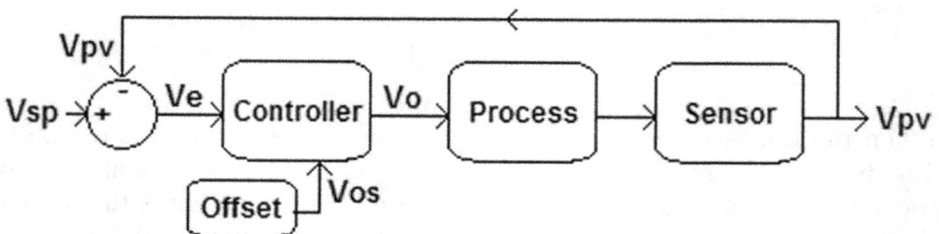

The "controller" is the component that reacts to the system's error and applies corrective action to the controlled process. Process examples include heaters, motors, valves, and power supplies. The "sensor" monitors the process. Its output is the process variable, **Vpv**. This may represent temperature, speed, position, or many other variables.

An "offset" may be needed to operate the process at zero error. Suppose a controller output of 3 volts is needed to maintain an oven at a set point temperature of 400 degrees. The error voltage will be zero since the oven temperature equals the set point temperature. An offset of 3 volts needs to be added to the error voltage to keep the oven temperature at the set point value.

Two Point On-Off Control System

Temperature control is a typical example of an on-off control system. In its simplest form, the heat is turned on when the temperature is below the set point, and turned off when the temperature is above the set point. The problem with this method is that the heater is turned on and off too often, which shortens the life of the control systems components.

A two point on-off controller is more common. For example, if the desired temperature is 70^0, the controller may be designed to turn off when the temperature exceeds 71^0, and turn on when the temperature drops below 69^0. The controller has hysteresis, similar to a Schmitt trigger circuit.

A graph of the temperature as a function of time of a two-point temperature controller is shown below. Notice the "over-shoot" above 71^0 and "under-shoot below 69^0.

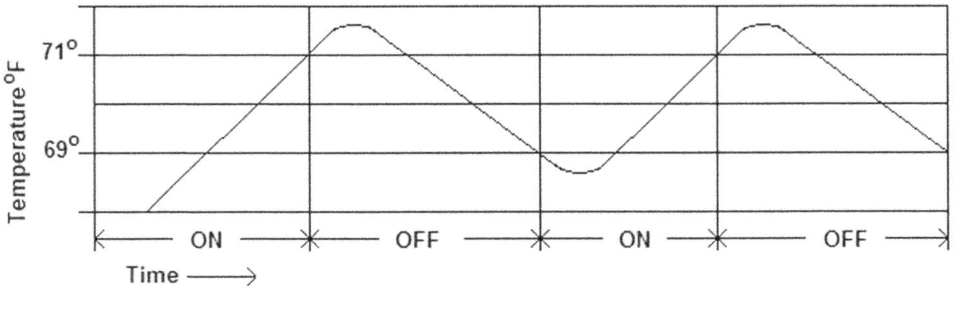

On-Off Control Circuits

An analog comparator can be used to sense if an input voltage is above or below the set point voltage. For example, the output of the circuit on the right will be zero volts if **Vpv** is above **Vsp**, and about 8 volts if **Vpv** is below **Vsp**.

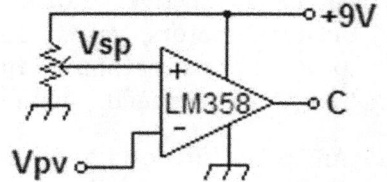

Two comparators and one SR latch are used in the circuit below to implement a two point controller. V_{LS} is the low set point voltage and V_{HS} is the high set point voltage. The output of the controller is the Q output of the latch.

If Q controls the power applied to a heater and **Vpv** is a voltage representing temperature, the heater will be on when **Vpv** is less than the low set point. The heater will not turn off until **Vpv** is higher than the high set point. It will then stay off until **Vpv** drops below the low set point. Refer to the table below.

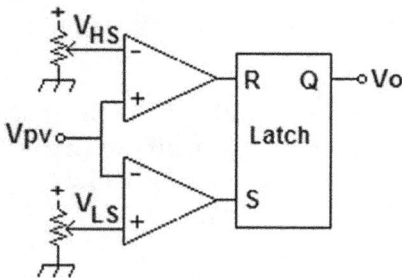

Process Variable	SR Latch Table		
Vpv	S	R	Q
Vpv < V$_{LS}$	1	0	1
V$_{LS}$ < Vpv < V$_{HS}$	0	0	no change
Vpv > V$_{HS}$	0	1	0

A two-point controller can be implemented with an op-amp Schmitt trigger circuit as shown on the right. **Vo** will be high **Vpv** is less than V_T, and **Vo** will be low when **Vpv** is greater than V_T. About 8 volts and 0 volts respectively for an LM358.

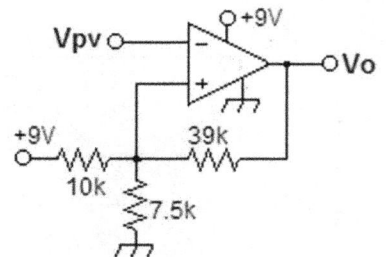

The "trip voltage" V_T depends on the output voltage **Vo**. Its value is designated as V_{UT}, the upper trip voltage, when **Vo** is high and V_{LT}, the lower trip voltage, when **Vo** is low.

Refer to an op-amp textbook to review the operating principles of the Schmitt trigger. The resistor values in the diagram above produce a V_{LT} of 3.4 volts and a V_{UT} of 3.7 volts.

The response of a controller using this circuit is shown below. **Vpv** is the output of a temperature sensor with a transfer function of 100mV/°C (3.4 volts = 34 °C and 3.7 volts = 37 °C.

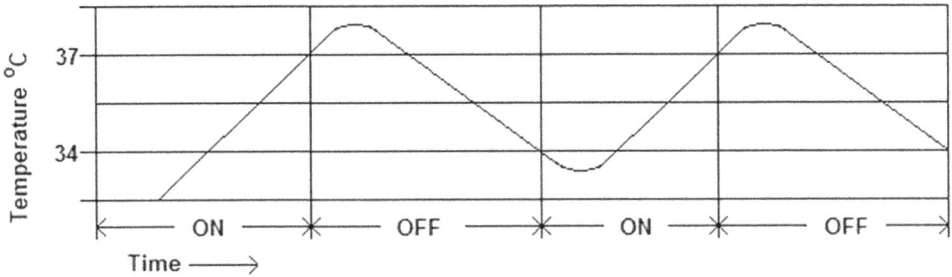

A block diagram of a two-point control system is presented below. The controlled process for a temperature controller involves turning a heater on and off. A temperature sensor is used to monitor the temperature and its output is compared to the low and high set points. The controller applies power to the heater when needed.

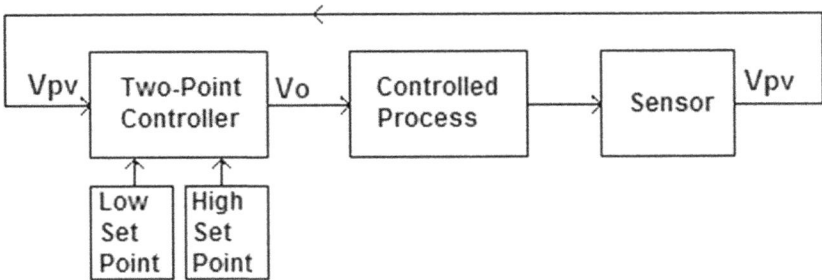

Thyristors, controlled by the controller's output voltage, Vo, are typically used to turn the heater power on and off. The temperature sensor may be a semiconductor, thermistor, resistance wire, or thermocouple type, depending on the required temperature range.

The experiment and simulation exercises to follow use an RC network to simulate a real process. They focus on control system principles and require a minimal number of parts. The process variable, **Vpv**, is the capacitor voltage. In effect, the controlled process is a voltage source. There are experiments in the second half of this book which involve heaters, motors, and sensor circuits.

Simulation of a two-point Controller

LTspice IV from Linear Technology may be used to simulate control system circuits. The information presented is intended to provide example simulations. You can download a manual and a "getting started guide" as well as using "help" to learn how to use the program, if necessary.

Connect the circuit on the right using the part "LT1006" from the "Opamps" library.

Double click on the capacitor to open its property editor and set initial condition to 0V (IC = 0).

Set the analysis type to "Transient".

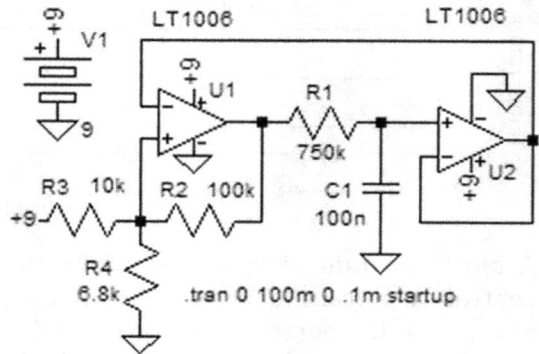

Set the "Stop Time" to 100mS and the "Maximum Time Step" to 0.1mS. Check the box "Start external supply voltage at 0V". Run the simulation. Probe the output of U2.

When the output of U1 is high (8 volts), the capacitor C1 will charge. U2 buffers the capacitor voltage and applies it to the U1, the Schmitt trigger. When the voltage reaches the upper trip level of U1, the out put of U1 will go low (0 volts). The capacitor will discharge until its voltage reaches the lower trip level of U1, and the cycle will repeat.

The graph on the right shows the simulation result for the output of U2.. Imagine that the RC network of R1 and C1 represents an oven and the capacitor voltage represents the oven temperature in units of 100mV / °C.

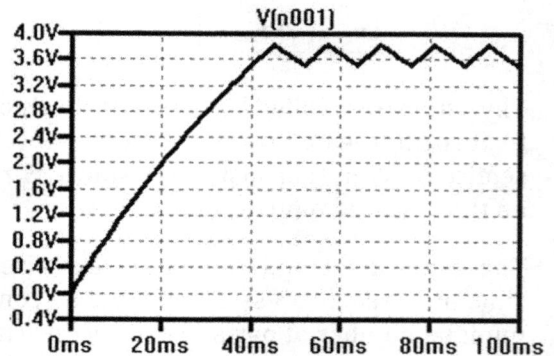

The graph shows that it takes about 50mS for the oven to go from 0°C to about 38°C.

The oven temperature cycles between 35°C and 38°C with a period of about 10mS.

34

Node voltage equations can be used to calculate the Schmitt trigger's trip levels:

$$\frac{V_{UT}-9}{R3}+\frac{V_{UT}}{R4}+\frac{V_{UT}-8}{R2}=0 \quad \text{and} \quad \frac{V_{LT}-9}{R3}+\frac{V_{LT}}{R4}+\frac{V_{LT}}{R2}=0$$

V_{UT} is the upper trip level and V_{LT} is the lower trip level. Using the values of R1, R2, and R3 in the simulation above results in V_{UT} = 3.8 volts and V_{LT} = 3.5 volts.

Experiment with the values of R2, R3, and R4. Increasing the value of R2 will decrease the amount of hysteresis (trip levels closer together). The average capacitor voltage can be varied with R3 and R4. Experiment by keeping the value of R3 + R4 about the same, but vary their ratio.

Experiment 5: Two Point Control System

This lab exercise uses a Schmitt trigger circuit as a two-point controller. Its output voltage, **Vo**, is applied to an RC network that simulates a process such as an oven heater. The capacitor voltage simulates the oven temperature.

When **Vo** is below the Schmitt trigger's "low set point", **Vo** is about 8 volts and the capacitor charges. When **Vo** goes above the "high set point", **Vo** is about 0 volts and the capacitor discharges. Feedback of the capacitor voltage, **Vpv**, to the control circuit keeps Vpv between the low and high set points.

Parts and Equipment Required

Oscilloscope, DMM,   Power Supply: ±9 volts.
Resistors: three 10k, 6.8k, 100k, 150k, 390k, all ¼W, 5%.
IC: LM358, Capacitors: 100μF, 470nF.

Procedure

1. Measure and record the values of R1, R2, and R3.

    R1 _____   R2 _____   R3 _____

2. Connect the circuit on the right. Apply power. Connect a DMM set to DC volts to terminal **Vpv** to monitor the capacitor voltage.

    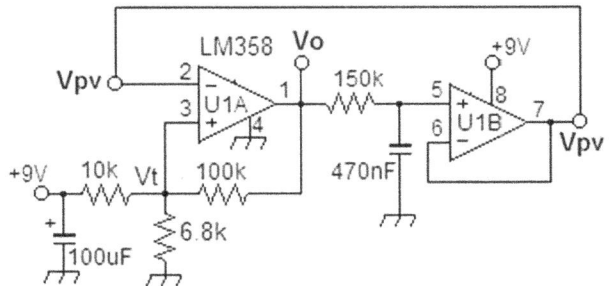

3. Connect oscilloscope to channel 1 to **Vo** and channel 2 to **Vpv**. Set trigger to channel 1.

    Set both channels to 1 volt per division and DC coupling. Set the zero reference for both channels to the bottom of the screen. Set the time base to 5mS per division.

    Your oscilloscope display should be similar to the one on the right.

    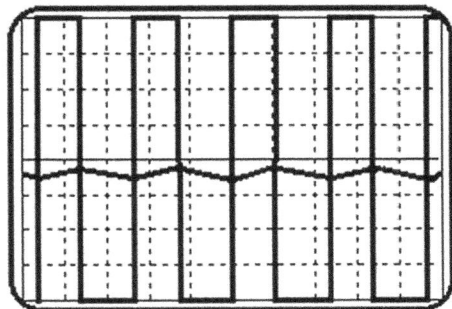

The display shows that **Vo** goes low when the triangle wave reaches about 3.8 volts on the positive slope and goes high when the triangle wave goes below about 3.5 volts.

4.   The "process variable" of this control system is the capacitor voltage represented by the voltage **Vpv**, which is applied to the trigger input of the controller.

5.   Measure and record the values of the upper and lower peaks of the triangle wave, **Vpv**.

      VpvUPPER _____   **Vpv**LOWER _____

6.   Measure the time intervals of the positive and negative slopes of Vpv in milliseconds. Reduce the oscilloscope's time per division to increase accuracy.

      VpvPOS _____   **Vpv**NEG _____

7.   Record the average value of **Vpv** from step 2.

      VpvAVE _____

8.   Connect a 390k ohm resistor in parallel with the capacitor. Repeat steps 5, 6, and 7.

      VpvUPPER _____   **Vpv**LOWER _____

      VpvPOS _____   **Vpv**NEG _____

      VpvAVE _____

Questions and Analysis

1.   Calculate the upper and lower trip levels of the Schmitt trigger using your measured resistor values. Compare results to your procedure step 5 and step 8 measurements. Express the differences from the calculated values in percent.

2.   Calculate the duty cycle of Vo from procedure step 6 and step 8 measurements. Explain the difference between step 6 and step 8 duty cycles due to the addition of the 390k ohm resistor.

3. Calculate the average value of Vpv from your oscilloscope measurements for procedure steps 5 and 8. Compare these values to the values measured by the DMM.

4. Simulate the circuit using your measured values of R1, R2, and R3. Compare the simulated results to your calculated results.

5. Project suggestion: Design a two-point controller using two comparators and an RS latch. Simulate the circuit. Build and test the circuit. Write a report.

Chapter 4: Proportional Mode Control System

The amount of corrective action applied by a proportional mode controller to a process is proportional to how far the process variable is from the set point. This allows the control system to reach the set point faster and reduces the possibility of overshoot and undershoot.

An example of a proportional mode control system is presented in the block diagram below. This is basically a voltage source whose output voltage can be varied by the set point setting.

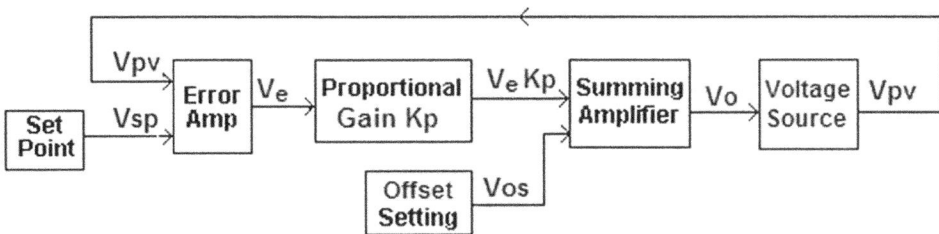

The output voltage, **Vpv**, of the voltage source is the process variable. **Vpv** is compared to the set point voltage, **Vsp** by a differential amplifier with a voltage gain of one. The difference is the error voltage which can be expressed as: **Ve = Vsp – Vpv.**

Ve is amplified by an operational amplifier with a gain of Kp. This amplified error voltage controls the voltage source output voltage, **Vpv**, in such a way as to reduce the error voltage (negative feedback). The output voltage of the controller, **Vo**, can be expressed as: **Vo = Ve·Kp + Vos.**

Study the block diagram above. When the error voltage is zero, the controller output voltage, **Vo**, will be the offset voltage, **Vos**. Typically **Vos** is set so that **Vpv = Vsp** at a desired resting or equilibrium value. For this one setting of **Vsp** the error voltage will be zero.

When **Vsp** is changed, the voltage needed to drive **Vpv** to the new setting is the amplified error voltage, **Ve·Kp**. Increasing the value of Kp will reduce the amount of error voltage needed so that the new setting will be approached with less error.

Proportional Band

The range of error voltage which produces an output that is proportional to the error is called the "proportional band".

The graph on the right is for a system whose process variable, **Vpv**, can vary from 0 to 8 volts.

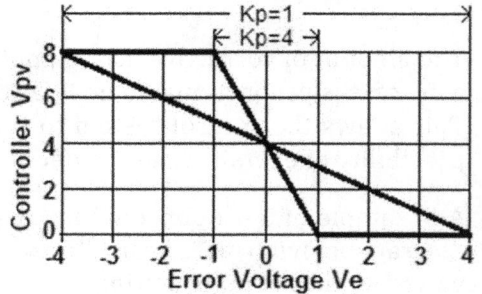

When **Vsp** = 4 volts, **Vpv** = 4 volts and **Ve** = 0.
When Kp = 1, the proportional band is ±4 volts.
When Kp = 4, the proportional band is ±1 volt.

When the error voltage is outside of the proportional band, the process variable will be the output saturation voltage, either 0 volts or 8 volts.

The graphs below show the control systems response for the voltages **Ve** and **Vpv** as the set point voltage, **Vsp**, is varied from 1.0 volts to 5.0 volts for proportional gains of one and four. Equilibrium value of **Vpv** is set to 3.0 volts. Ideally, **Vpv** should coincide with the dashed line.

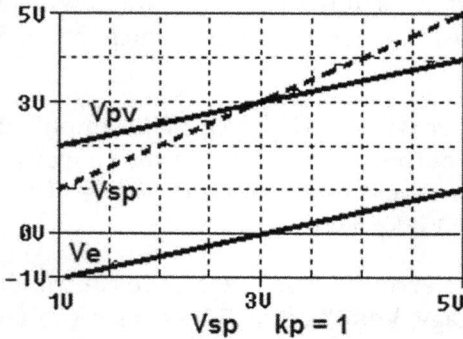

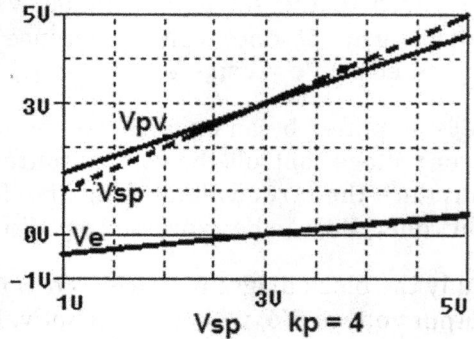

Steady State Response

Given Vo = Ve·Kp + Vos and that the steady state value of Vpv = Vo, the steady state value of Vpv is calculated below:

$$\mathbf{Vpv} = \mathbf{Vo} = Kp(\mathbf{Vsp} - \mathbf{Vpv}) + \mathbf{Vos} \quad \Rightarrow \quad \mathbf{Vpv} = \left(\frac{Kp}{1+Kp}\right)\mathbf{Vsp} + \left(\frac{1}{1+Kp}\right)\mathbf{Vos}$$

The result shows that as Kp approaches infinity, Vpv approaches Vsp.

42

Transient Response

The transient response of a control system occurs from the time a new set point is initiated to the time the system reaches and settles on the new set point value.

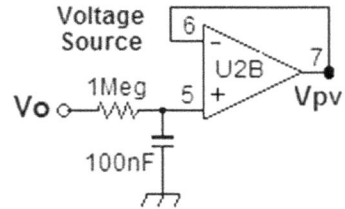

This exercise uses the voltage source shown on the right for the controlled process. An RC network with a 0.1 second time constant introduces (simulates) a time delay to the process.

Time or frequency dependant transfer functions must be used to calculate the transient response of the control system. In this case, only the time or frequency dependence of the voltage source needs to be considered because it has a much slower time response than the rest of the system. That is, the rise and fall times of the other system components can be neglected.

The frequency response of the control system can be easily calculated as the frequency response of the voltage source (given, in this case, that the cutoff frequency of this voltage source is much lower than the cutoff frequency of the other system components). However, the transient response calculation is not as easy because the system's response to a step change in the set point needs to be determined.

Algebra alone is not sufficient for this calculation. If you are not familiar with the math, you may skip down to the final result.

The control system's transient response is calculated below using the Laplace transform method. A simplified block diagram is shown on the right.

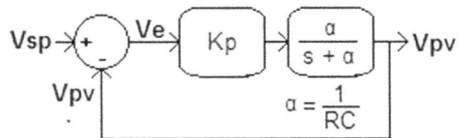

$$\mathbf{Vpv} = Kp(\mathbf{Vsp} - \mathbf{Vpv})\frac{\alpha}{s+\alpha} \quad \Rightarrow \quad \mathbf{Vpv}\left(1 + \frac{\alpha}{s+\alpha}Kp\right) = \frac{\alpha}{s+\alpha}Kp\mathbf{Vsp}$$

$$\mathbf{Vpv} = \frac{\dfrac{\alpha}{s+\alpha}Kp\mathbf{Vsp}}{\left(1 + \dfrac{\alpha}{s+\alpha}Kp\right)} = \frac{\alpha Kp}{s+\alpha(1+Kp)}\mathbf{Vsp}$$

$$\mathbf{Vsp} = \frac{1}{s}, \text{ a unit step, then, } \quad \mathbf{Vpv} = \frac{\alpha Kp}{s(s+\alpha(1+Kp))}$$

Expanding result by partial fractions and transforming to the time domain yields:

$$Vpv = \left(\frac{Kp}{1+Kp}\right)\left(1-e^{-\alpha(1+Kp)t}\right)$$

This result is for a unit step with no offset.

Final result (A is the step size in volts, Vos is the offset voltage, and t is time in seconds):

$$Vpv = \left(\frac{AKp}{1+Kp}\right)\left(1-e^{-\alpha(1+Kp)t}\right)+\textbf{Vos.} \quad \textbf{At } (t=\infty), \ Vpv = \left(\frac{AKp}{1+Kp}\right)+\textbf{Vos.}$$

This result shows that as the value of Kp increases the step size approaches A and the response time (rise and fall times) decreases. This is shown by the simulation results below.

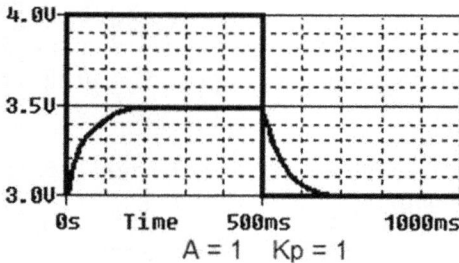

A = 1   Kp = 1

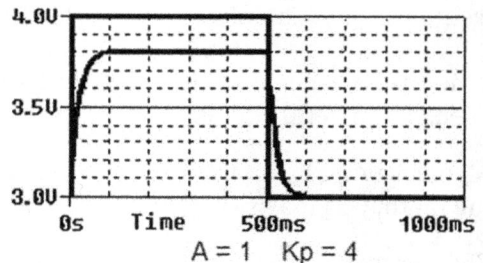

A = 1   Kp = 4

Simulation results above show the control system's transient response in going from the equilibrium set point of 3.0 volts to a set point of 4.0 volts and back to 3.0 volts. A Kp of 1.0 results in a steady state error of 0.5 volts when going to 4.0 volts. A Kp of 4 results in a steady state error of 0.2 volts when going to 4.0 volts and a much shorter rise times, fall times, and settling times.

Simulation of Proportional Control

Study the LTspice circuit below. U1A is a unity gain differential amplifier Its output is the error voltage, **Vsp** – **Vpv**.

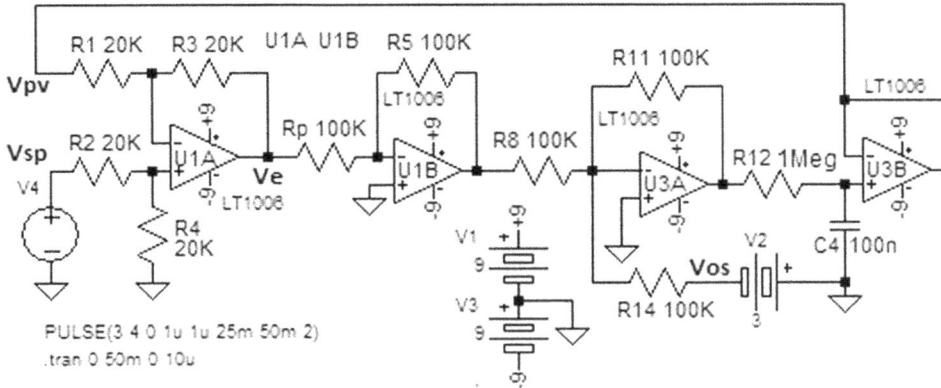

U1B sets the proportional gain: Kp = R5/RP. U3A sums the proportional output, **Ve**Kp and the offset voltage, **Vos.** R12 and C4 add a 100mS time constant to the process.

Set the simulation to "Transient' with a stop time of 50mS. Plot the voltages **Vsp** and **Vpv**.

Study the graph on below. The response for the first 250mS shows the capacitor charging from 0.0 volts to 3.5 volts. During this time the error voltage changes from 4.0 volts to 0.5 volts.

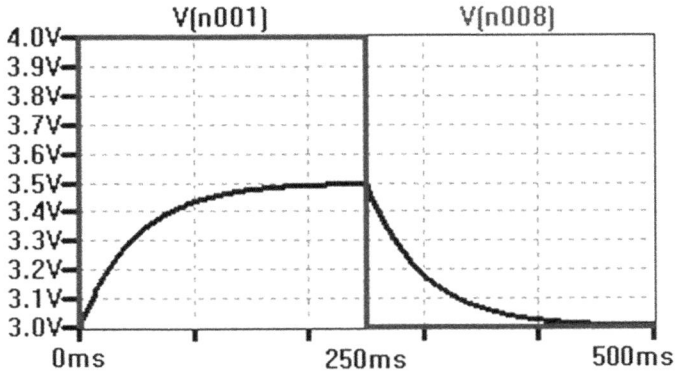

The next 250mS interval show that the error voltage settles on 0.0 volts when the set point voltage is the equilibrium value of 3.0 volts.

Experiment with the simulation by changing the proportional gain, Kp, the set point, **Vsp**, and the offset voltage, **Vos**. The result below is for Kp = 4.

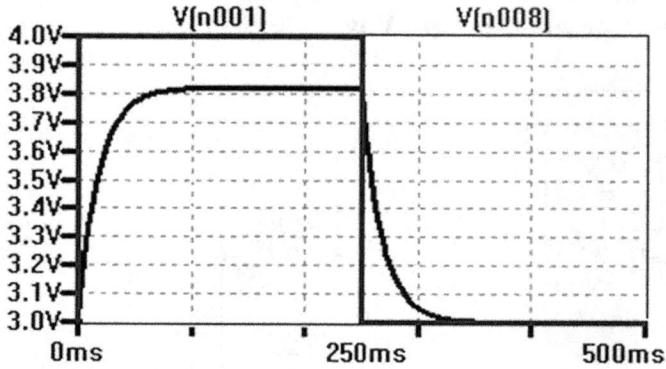

Compare the simulation results to the results predicted by the equation:

$$\mathbf{Vpv} = \left(\frac{A Kp}{1+Kp} \right) \left(1 - e^{-\alpha(1+Kp)t} \right) + \mathbf{Vos}. \qquad Kp = \frac{100k}{Rp}$$

Experiment 6: Proportional Mode Control System

The PID-X1 controller board may be used in this exercise to implement a proportional control system. The controlled process is a voltage source whose set point voltage is switched between 3 volts and 4 volts by a square wave. The transient and steady state response of a proportional control system to a change of set point will be measured for several values of amplifier gain, Kp. The responses will be compared to simulations and calculations. This lab experiment demonstrates the control system concepts of proportional band, proportional error, response time, and settling time.

Parts and Equipment Required

Oscilloscope, DMM, Function Generator.
Power Supply: ±9 to ±12 volts.
Controller board PID-X1 (or build the circuit on a breadboard).
Resistors: 10k, 100k, 1Meg, all ¼W, 5%.

Procedure Part 1: Steady State Response

1. The control circuit schematic diagram is given below. If using the PID-X1 control board, make the connections as shown below. In addition, connect a 1 megohm resistor from pin I (integrator input) to ground. Refer to the picture on the next page.

    If you are not using the PID-X1 control board, layout and build the circuit as carefully as possible. Keep the number of wires to the absolute minimum. Observe the numbers on the ICs. For example, U1A and U1B are in the same package.

    Observe the polarities on the 100µF electrolytic capacitors.

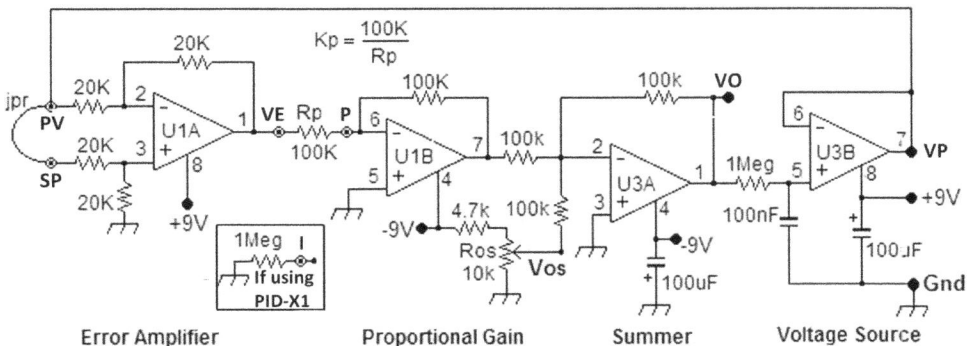

PID-X1 note: **Vpv = PV, Vsp = SP, Ve = VE, Vos = VO** when **Ve** = 0.

Layout the circuit in the same order as the schematic, U1A error amplifier circuit on the far left and the transistor on the far right. It is important that you can easily identify each functional block of the circuit on the breadboard.

A jumper wire "jpr" is connected between **PV** and **SP** temporarily to set the equilibrium value of **VP** (the zero error value of **VP**).

1b.   If you are using the PID-X1, plug it into your breadboard and make the connections shown below.

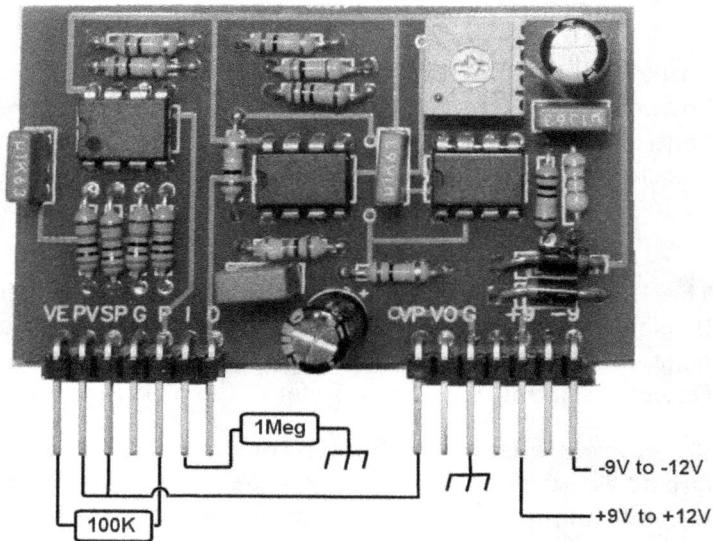

2.   Apply power to the circuit.   Check that the voltage **VE** is zero.   It should be very close to zero since **PV = SP** (due to the jumper wire).

Adjust the potentiometer, Ros, to set **VO** to exactly 3.0 volts.   Use a DMM to make the measurement.   Measure the resulting value of **VP** and check that it is equal to **VO** (3V).

3.   Turn off power.   Remove the jumper between **PV** and **SP**.   A variable voltage source needs to be connected to **SP**.   This may be a separate power supply or use a potentiometer connected to +9V (or up to +12V) and ground as shown on the right.

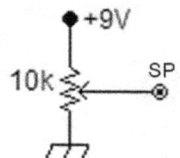

4. Set **SP** to 3.0 volts. Verify that **VE** is close to 0 (less than 0.1 volts). Set **SP** to the following voltages and each time measure **VE** and **PV**: 0, 2, 3, 4, 5, 6, 7, 8 Record the measurements in the table below.

SP(Vsp)	0.0	1.0	2.0	3.0	4.0	5.0	6.0	7.0	8.0
PV(Vpv)									
VE(Ve)									

5. Remove the potentiometer from **SP**.

Procedure Part 2: Transient Response

1. Set the function generator to produce a 1 volt peak-to-peak, 1Hz, square wave with a 3.5 volt offset (so it goes between 3.0 volts and 4.0 volts). Connect the function generator to **SP**. Turn on power to control circuit.

Connect the oscilloscope channel 1 to **SP** and Channel 2 to **PV**. Set vertical inputs to DC coupling and 200mV/Div. Set horizontal to 100mS/Div and Trigger on channel 1. Set vertical position to obtain a display similar to that shown on the right.

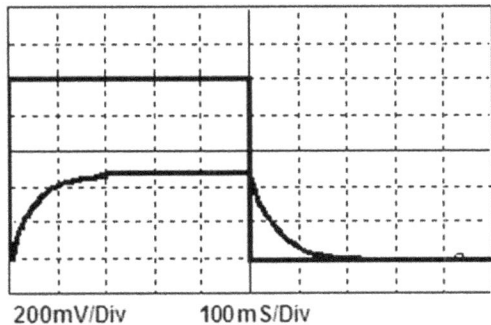

200mV/Div   100 mS/Div

3. From your oscilloscope display, determine the steady state value, rise time, and settling time of **PV (Vpv)**. Record below.

Steady State _____ volts   Rise Time _____ mS

Settling Time _____ mS

4. Replace Rp in the controller circuit with a 10k resistor to set Kp to 10. Repeat steps 1 through 3 and record results below. You should reduce the oscilloscopes time per division to measure the rise time of **PV** more accurately.

Steady State _____ volts   Rise Time _____ mS

Settling Time _____ mS

Analysis Part 1

1. Calculate the expected steady state errors for Kp =1, Vsp = 2 volts, **SP** = 4 volts, and **SP** = 6 volts (should have zero error for **SP** = 3 volts). Compare calculations to measured results. Express the percent difference between the measurements and calculations.

2. Calculate the expected steady state errors for Kp =10, Vsp = 2 volts, **SP** = 4 volts, and **SP** = 6 volts (should have zero error for **SP** = 3 volts). Compare calculations to measured results. Express the percent difference between the measurements and   calculations.

3. Calculate the controller's proportional band for the set point of 3.0 volts with Kp = 1 and   Kp = 10. Assume that the op-amp saturation voltage is 1 volts less than the supply voltage and that the diode saturation voltage is -0.7 volts.

Analysis Part 2

1. Calculate the time constant of the RC filter in the voltage source (100nF capacitor and 1Meg resistor). Calculate α (this is also called the "decay rate" and has units of "nepers per second"). Compare the rise time of **PV** of the control system with Kp = 1 to the calculated time constant.

2. Compare the measured results of procedure part 2, steps 3 and 4, to calculated results. Use the equation below with the appropriate values of Kp, **Vos**, and α (**Vos** = **VO** when **VE** = 0).

$$\mathbf{Vpv} = \left(\frac{AKp}{1+Kp} \right) \left(1 - e^{-\alpha(1+Kp)t} \right) + \mathbf{Vos.}$$

3. Calculate the effect of connecting a 220k ohm resistor in parallel with the 100nF capacitor. What happens to α? What happens to **PV** when **Vos** is 3.0 volts and **SP**, the set point. is 3.0 volts. What happens to **PV** when **Vos** is 3.0 volts and the set point is 4.0 volts.

Chapter 5: Proportional Plus Integral Mode Control

The proportional error problem of the proportional controller is solved by the addition of an integrator circuit. Mathematically, the process of integration is the process of finding the area under a curve. Consider the graphs below where the vertical axis is volts and the horizontal axis is time.

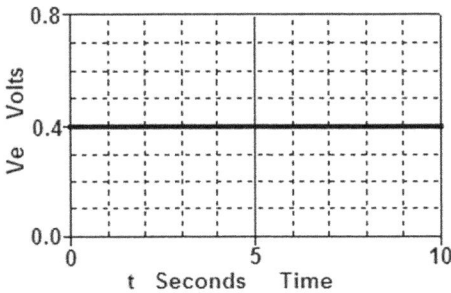

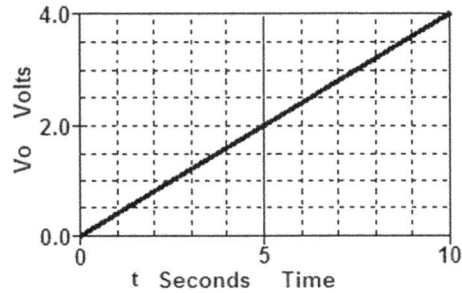

The graph above on the left shows the input voltage applied to an integrator, Ve. The graph on the right shows the resulting output voltage, **Vo**. The input voltage is a constant 0.4 volts. The output voltage increases with time as the product of **Ve** and **t**. **Vo = Ve·t**.

This shows that as long as there is an error voltage, **Ve**, at the input of the integrator, the integrator output voltage, **Vo**, will continue to increase. In a feedback control system the output voltage is fed back to the controller input, causing the error voltage to decrease until it is zero. Therefore the error voltage will not be constant with time, but will continually decrease. As the error voltage decreases the integrator output voltage will change more slowly until it reaches the constant value required to produce a zero error voltage.

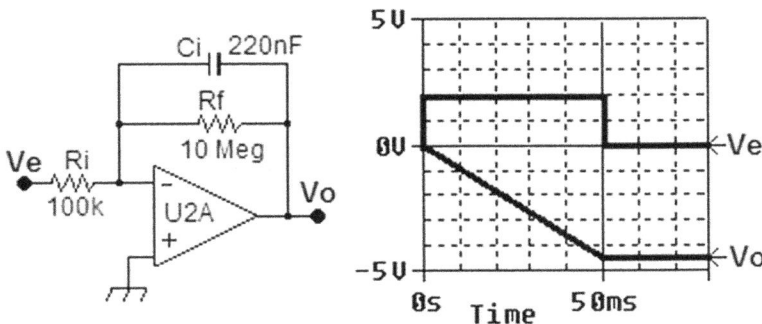

An op-amp integrator and a graph of its input and output voltages are shown above. An input voltage of 2 volts is applied for 50mS producing a current of

20μA through the 100k resistor. Almost all of this current charges the capacitor. The output voltage, **Vo**, is negative, reaching -5 volts in 50mS.

When the input voltage goes to 0 after 50mS, the capacitor current goes to 0, but the capacitor remains charged, and **Vo** remains at -5 volts. When the input voltage is constant and the initial capacitor voltage is zero, the output voltage can be expressed mathematically as:

$$\textbf{Vo} = \frac{-1}{Ri \cdot Ci} \textbf{Ve} \cdot t = -90.9 \cdot t, \quad \text{for } t < 50mS \qquad \textbf{Vo} = 4.55 \text{volts} \quad \text{for } t > 50mS$$

When Ve is not constant and the capacitor has an initial voltage of Vi at t = 0, the equation for **Vo** is:

$$\textbf{Vo} = -Ki \int_0^t \textbf{Ve} \cdot d\tau + Vi \quad \text{where } Ki = \frac{1}{Ri \cdot Ci}. \quad \text{Ki is the integrator gain.}$$

The integrator gain, Ki, and the proportional gain, Kp, are very important control system parameters. A control system is a "tuned" to obtain an optimum response by adjusting these parameters. A block diagram of a "PI" control system is shown below. The integrator's output is summed with the proportional output and with the offset voltage.

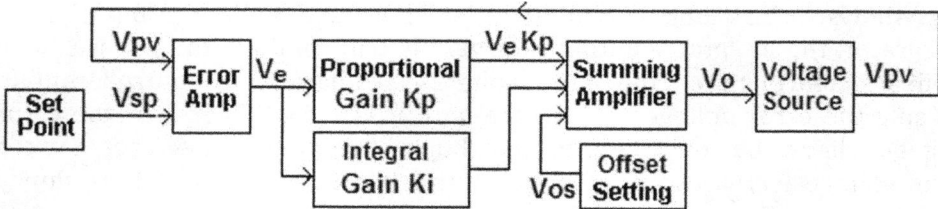

The equation for the controller's output voltage is given below.

$$\textbf{Vo} = Kp \cdot \textbf{Ve} + Ki \int_0^t \textbf{Ve} \cdot d\tau + Vi + \textbf{Vos}$$

Calculating the controller's output is complicated, however, much can be inferred about the response from the roots of the controller's transfer function.

Refer to the diagram on the right. There are two time constants in this circuit: α = 1/RC, the reciprocal of the time constant of the process, and RiCi, the time constant of the integrator.

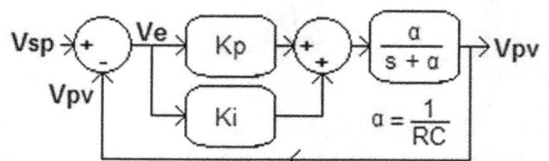

The s-domain transfer function is given below. Its denominator roots predict the type of response.

$$H_f = \frac{\text{Vpv}}{\text{Vsp}} = \frac{\alpha Kp\left[s+(Ki/Kp)\right]}{s^2 + s\alpha(1+Kp) + \alpha Ki}.$$

Denominator roots: $\dfrac{-\alpha(1+Kp)}{2} \pm \dfrac{\sqrt{\left[\alpha^2(1+Kp)\right]^2 - 4\alpha Ki}}{2}$

An under-damped response occurs when the radical component becomes imaginary and the roots are complex. The result is oscillation, which is undesirable. A critically damped response occurs when the radical component is zero. However a small change in the control system could produce oscillation. Therefore an over-damped response is desirable. In this case:

$$\alpha^2(1+Kp)^2 > 4\alpha Ki \Rightarrow Ki < \frac{\alpha(1+Kp)^2}{4}. \quad \text{Ki should be less than } \frac{\alpha(1+Kp)^2}{4}.$$

Simulation of a Proportional Plus Integral Control System

The LTspice circuit below uses the "LT1006" from the "Opamps" library. It is similar to the more common parts, LM358 and LM324. U1A is a unity gain differential amplifier. Its output is the error voltage. This error voltage is applied to the proportional amplifier, U1B, and the integrator, U2A.

The outputs of U1B and U2A are summed by a unity gain summing amplifier, U3A. Resistor R14 sums the offset voltage, V2. Note that V2 is negative because U3A is an inverting amplifier and the desired offset voltage is positive.

Although the gains of U1B (Kp) and U2A (Ki) are negative, their net value is positive because of the inverting op-amp, U3. The process variable, Vpv, connects to the inverting input of the differential amplifier, providing negative feedback.

The output of U3A is applied to a low pass filter, R12 and C4, with a time constant of 100mS to simulate the control system process (heater, motor, etc.). This may be modified as needed. U3B is a unity gain buffer.

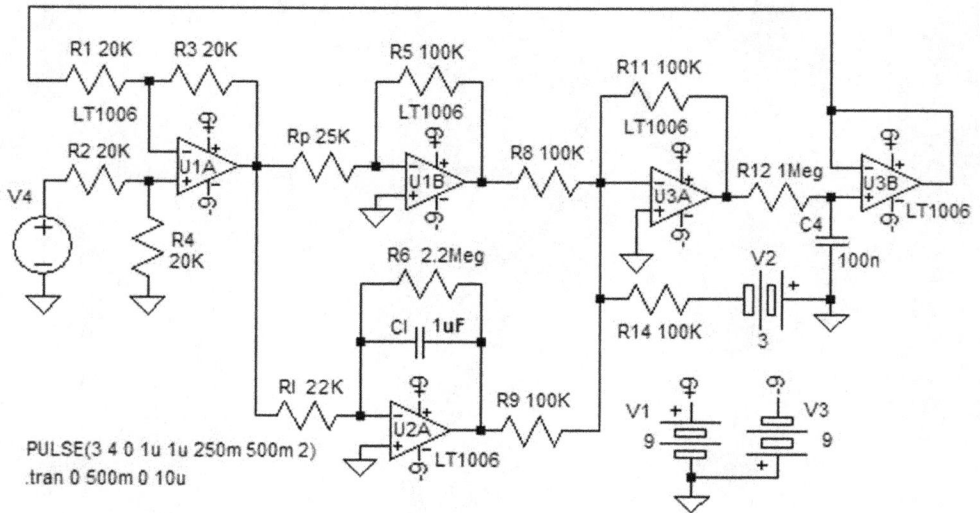

Proportional mode simulation is done first by disconnecting R9. This is so that the controller's response without the integrator can be compared to its response with the integrator.

Transient analysis is used with a "Stop Time" of 500mS and a "Maximum Timestep" of 100uS.

Proportional mode results are shown on the right for the two inputs to the differential amplifier, **Vpv** and **Vsp**.

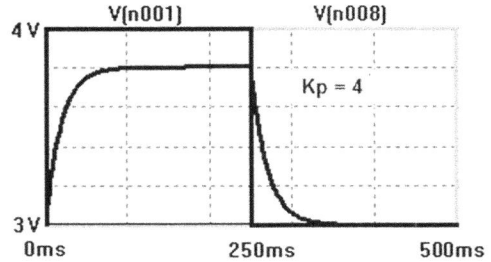

Add integral mode by reconnecting R9. The values of Ri and C2 result in an integral gain, Ki, of 45. Transient analysis is used with a "Stop Time" of 500mS and a "Maximum Timestep" of 100uS.

The results are shown on the next page for the two inputs to the differential amplifier, **Vpv** and **Vsp**. Both responses are for a Kp of 4. Adding the integrator nearly eliminates the steady state error. Increasing the integral gain decreases the rise time, but too much integral gain can produce oscillation and increase the settling time.

The response on the left is for an integrator gain of 45 (Ri = 22k). It shows a steady state error of nearly zero with no over-shoot or under-shoot. The settling time is about 10mS. The response on the right is for an integral gain of 213. It shows an over-shoot and an under-shoot of about 0.2 volts and a settling time of about 20mS.

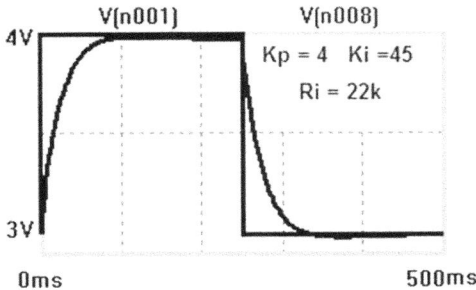

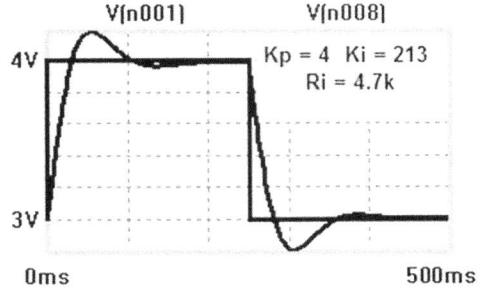

Experiment 7: Proportional Plus Integral Mode Control System

This lab exercise adds an op-amp integrator to the proportional mode control system. The controlled process is a voltage source whose set point voltage is switched between 3 volts and 4 volts by a square wave.

The transient and steady state response of a proportional plus integral control system to a change of set point will be measured for several values of proportional gain, Kp and integral gain, Ki. The responses will be compared to simulations. This lab experiment demonstrates the control system concepts of steady state error, response time, and settling time, overshoot, undershoot, and damping.

Parts and Equipment Required

Oscilloscope, DMM, Function Generator. Power Supply: ±9V to ±12V. Controller board PID-X1 (or build the circuit on a breadboard). Resistors: 4.7k, 20k, 22k, 47k, 100k, 1Meg, all ¼W, 5%.

Procedure Part 1: Proportional Mode Check

1. The control circuit schematic diagram is given below. If you are not using the PID-X1 plug-in, layout and build the circuit as carefully as possible. Keep the number of wires to the absolute minimum.

    Observe the numbers on the ICs. For example, U1A and U1B are in the same package. Observe the polarities on the 100µF electrolytic capacitors.

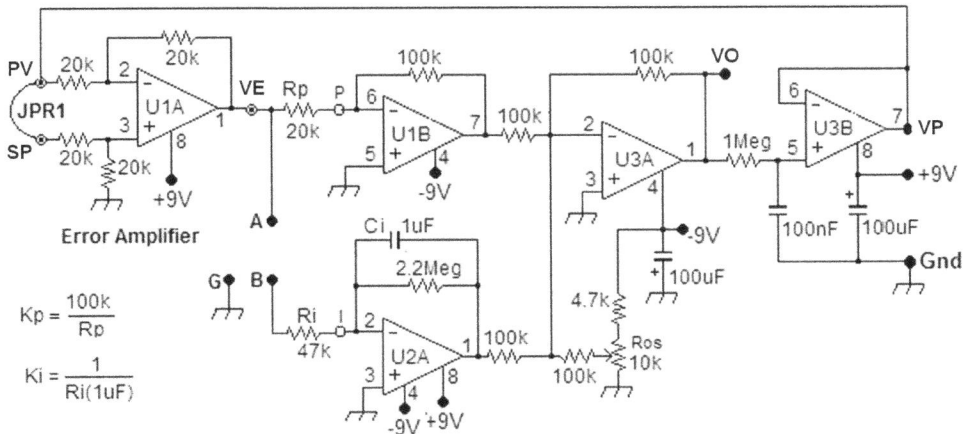

PID-X1 note: **Vpv = PV, Vsp = SP, Ve = VE, Vos = VO** when **Ve = 0**.

57

1b. If you are using the PID-X1, plug it into your breadboard and make the connections shown below.

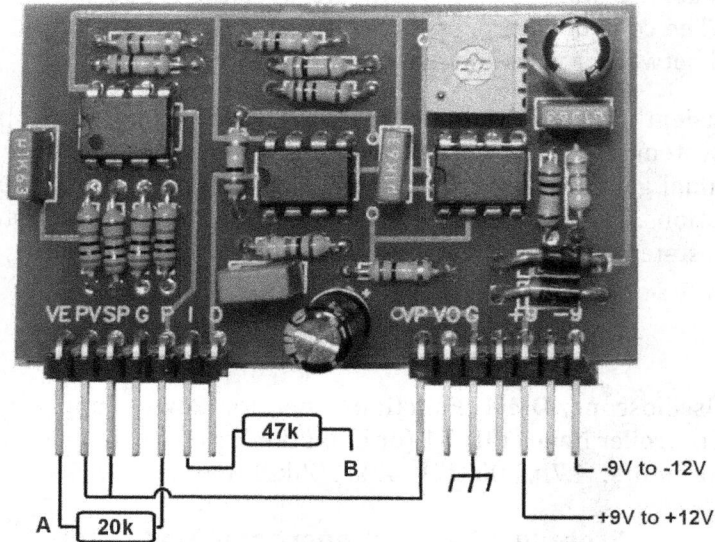

2. **JPR1** should be connected. Connect a 1Meg resistor between points **B** and **G** (ground). Note that R_P is 20k so that the proportional gain is 5. Apply plus and minus 9 volts to the circuit. Check that the voltage **VE** is zero. It should be very close to zero since **PV = SP** because of the jumper wire.

   Adjust the potentiometer, Ros, to set **PV** to exactly 3.0 volts. Use a DMM to make the measurement.

3. Remove the jumper between **PV** and **SP**.

4. Set the function generator to produce a 1 volt peak-to-peak, 1Hz, square wave with a 3.5 volt offset (so it goes between 3.0 volts and 4.0 volts). Connect the function generator to **SP**.

5. Connect oscilloscope channel 1 to **SP** and Channel 2 to **PV**. Set inputs to DC coupling and 200mV/Div. Set horizontal to 100mS/Div and Trigger on channel 1. Set vertical position controls to obtain a display similar to that on the right.

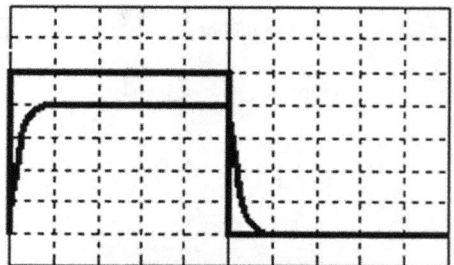

200mV/Div 100mS/Div

58

6. From your oscilloscope display, determine the steady state value, rise time, and settling time of **PV**. Record below for Rp = 20k. [Kp = 5 and Ki =0]

Steady State _____volts   Rise Time _____ mS

Settling Time _____mS

Procedure Part 2: Proportional Plus Integral Controller

1. Remove the 1Meg resistor between points **B** and **G**. Connect point **B** to **A**. Channel 1 should still be connected to **SP** and Channel 2 to **PV**. Vertical inputs set to DC coupling and 200mV/Div. Horizontal to 100mS/Div and Trigger on channel 1.

2. From your oscilloscope display, determine the steady state value, rise time, and settling time of **PV**.

   Record below for R_P = 20k and Ri = 100k. [Kp = 5, Ki = 21.3]

   Steady State _____volts   Rise Time _____ mS

   Settling Time _____mS

3. Replace the 100k resistor Ri in the integrator circuit with a 47k resistor. From your oscilloscope display, determine the steady state value, rise time, and settling time of **PV**. Record below for R_P = 20k and Ri = 47k. [Kp = 5, Ki = 45.3]

   Steady State _____volts   Rise Time _____ mS

   Settling Time _____mS

3. Replace the 22k resistor Ri in the integrator circuit with a 4.7k resistor. From your oscilloscope display, determine the steady state value, rise time, and settling time of **PV**. Record results for R_P = 20k and Ri = 10k. [Kp = 5, Ki = 213]

   Steady State _____volts   Rise Time _____ mS

   Settling Time _____mS

59

Analysis Part 1

1. Calculate the expected steady state error and rise time for experiment part 1 and compare results to the measurements. Express the percent difference between the measurements and calculations.

2. Calculate the controller's proportional band for experiment part 1.

Analysis Part 2

1. Simulate the circuit for R_P = 100k and Ri = 100k. Compare the measured and simulated steady state values, rise times, and settling times of **Vpv**.

2. Simulate the circuit for R_P = 20k and Ri = 47k. Compare the measured and simulated steady state values, rise times, and settling times of **Vpv**.

3. Simulate the circuit for R_P = 20k and Ri = 10k. Compare the measured and simulated steady state values, rise times, and settling times of **Vpv**.

4. Optional: Calculate the denominator roots for analysis step 1 and analysis step 3 using the result given below:

$$\text{Roots} = \frac{-\alpha(1+Kp)}{2} \pm \frac{\sqrt{\left[\alpha^2(1+Kp)\right]^2 - 4\alpha Ki}}{2}.$$

$$\alpha = \frac{1}{RC} = \frac{1}{(1Meg)(100nF)}.$$

Determine if the response is over-damped, critical-damped, or under-damped. If the response is under-damped the roots will be complex and can be written as:

$$\text{Roots} = \gamma \pm j\omega_d \qquad \gamma = \frac{-\alpha(1+Kp)}{2} \qquad \omega_d = \frac{\sqrt{\left[\alpha^2(1+Kp)\right]^2 - 4\alpha Ki}}{2}$$

γ: decay rate in nepers/sec. ω_d : damped frequency in radians/sec.

Chapter 6: PID Mode

A PID controller can be obtained by adding a differentiator to the PI controller. The differentiator is placed in parallel with the proportional and integral components here, although other arrangements are possible.

The mathematical analysis of a PID control system is beyond the scope of this presentation, however the response of the PID system can be estimated by understanding the response of the individual system components..

The magnitude of the derivative response is proportional to the rate of change of the controller's error voltage. The differentiator can improve a control system's response to rapid changes in the process variable. Mathematically, the process of differentiation is the process of finding the slope of a curve.

The ideal differentiator's contribution to the controller's output voltage is given by the equation:

$$vo = Kd \frac{d(Ve)}{dt}. \quad Kd = differentiator\ gain.$$

Consider the graphs on the right, where the vertical axis is Ve in volts and the horizontal axis is time in seconds.

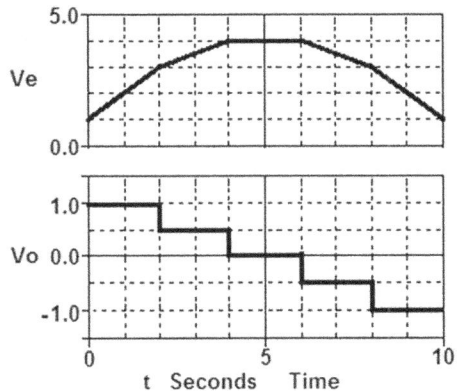

The slope (d(Ve)/dt) is held constant for intervals of 2 seconds.

Note that Vo is zero when the slope is zero and that Vo is negative when the slope is negative.

An op-amp differentiator is shown on the right. Although the gain of the differentiator, Kd, is given by the time constant of RdCd, the response of the circuit also depends on the value of the resistor Rx.

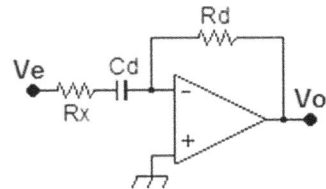

This is an inverting circuit so that Kd is actually negative, but this is compensated for by an inverting summing amplifier. The ideal response of this differentiator (Rx = 0) is given by:

$$\mathbf{Vo} = -Kd \frac{d(Ve)}{dt} \qquad Kd = (Rd)(Cd). \qquad Kd\ is\ the\ differentiator\ gain.$$

Kp, Ki, and Kd, are important PID control system parameters. A control system is a "tuned" to obtain an optimum response by adjusting these parameters.

The mathematical equation for an ideal parallel connected PID control system is given below.

$$\textbf{Vo} = Kp \cdot \textbf{Ve} + Ki\int_0^t \textbf{Ve} \cdot d\tau + Vi + Kd\frac{dVe}{dt} + \textbf{Vos}$$

A block diagram of a PID control system is shown below. The differentiator's output is summed with the integrator's output, the proportional output, and with the offset voltage.

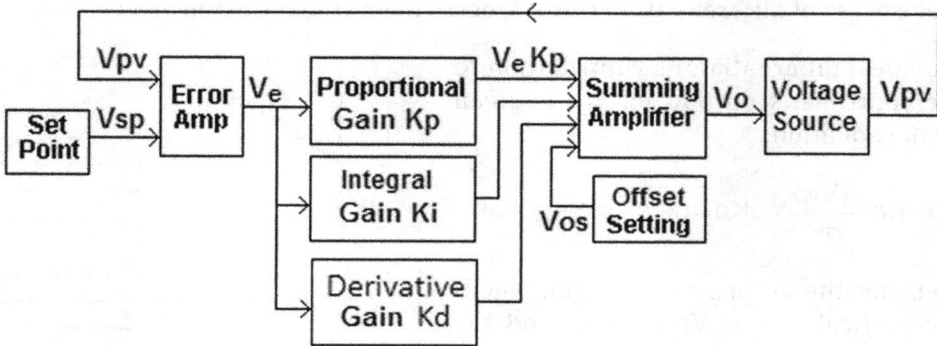

In the block diagram above the proportional, integral, and derivative components are driven by the same error voltage and the output of each component is summed by the summing amplifier. Other arrangements are possible. For example, the proportional and integral components could be connected in series.

The parallel form is chosen here because each component can be adjusted independently and it is also the typical arrangement discussed in electronic technology textbooks. The intention here is to introduce the basic components of a PID controller, and to demonstrate its operation by circuit simulation and experiment.

A PID control system is tuned for an optimum response by adjusting the gain of each system component, proportional, integral, and derivative. There are a variety of theoretical methods for doing this, but in this presentation tuning is done by repetitive simulation and experiment.

Simulation of a PID Control System

The LTspice circuit shown below uses the part "LT1006" from the "Opamps" library. U1A is a unity gain differential amplifier. Its output is the error voltage, Ve. This error voltage is applied to the proportional amplifier, U1B, integrator, U2A, and differentiator, U2B.

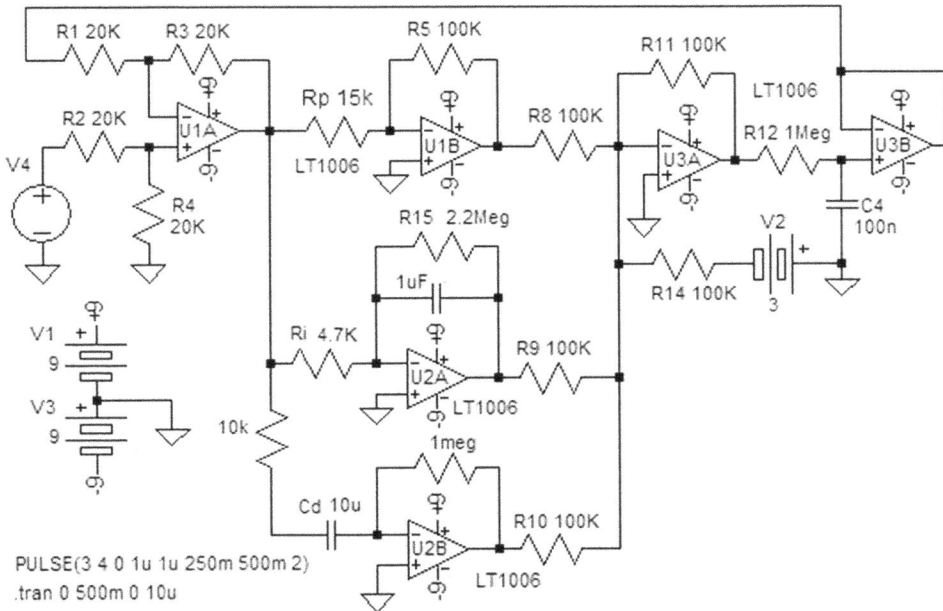

The outputs of U1B, U2A, and U2B are summed by a unity gain summing amplifier, U3A. Resistor R8 sums the offset voltage, V2. Note that V2 is negative because U3 is an inverting amplifier and the desired offset voltage is positive.

Although the gains of U1B (Kp), U2A (Ki), and U2B (Kd) are negative, their net value is positive because of the inverting op-amp, U3. The process variable, Vpv, connects to the inverting input of the differential amplifier, providing negative feedback.

The output of U3A is applied to a low pass filter, R12 and C4, with a time constant of 10mS to simulate the control system process (heater, motor, etc.). This may be modified as needed. U3B is a unity gain buffer

Proportional only simulation is done by removing R9 and R10. Proportional plus integral mode is done by keeping R9 in the circuit and removing R10. Proportional plus integral plus derivative (PID) mode simulation is done by keeping both R9 and R10 in the circuit. The derivative gain is set by Cd.

Transient analysis is used with a "Stop Time" of 500mS and a "Maximum Timestep" of 100uS.

The results are shown below for the two inputs to the differential amplifier, **Vpv** and **Vsp**. Both responses are for a Kp of 6.7(Rp = 15k) and Ki of 213 (Ri = 4.7k). The response below on the left is without the differentiator. The response on the right was obtained by adding a differentiator with a gain of 10 (Cd = 10µF).

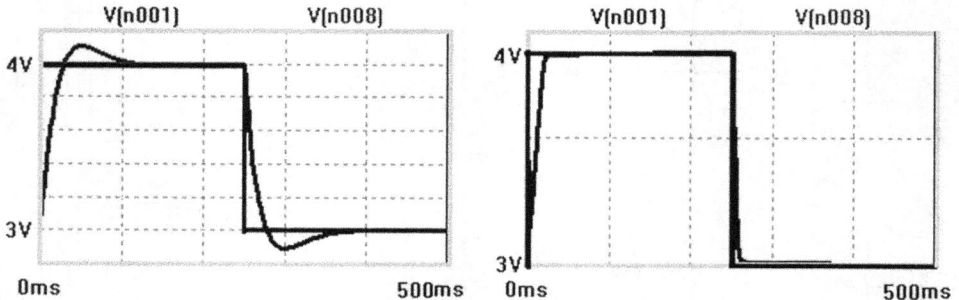

The mathematics behind tuning the PID controller is somewhat complex and will not be covered here. However, some observations may be made from this simulation.

1. The integrator's time constant is the reciprocal of its gain, which is 4.7mS in this case. Using the earlier result for the stability of a PI controller we see that the integral gain of 213 is too high, and the PI controller output will oscillate, as shown in the graph above on the left.

$$\alpha = \frac{1}{R12 \cdot C4} = 10. \qquad Ki < \frac{\alpha(1+Kp4)^2}{4} = \frac{10(1+6.8)^2}{4} = 152.$$

2. A differentiator is added to damp out the oscillations. The gain of the differentiator can be experimentally (using simulation) varied until the desired response is obtained. Here the differentiator's time constant is 10 seconds, 2130 times greater than the integrator's time constant of 0.0047sec.

3. Increasing the integrator gain until oscillation occurs shortens the rise time. Adding the differentiator damps out the oscillation and reduces the rise time further.

Experiment 8: PID Mode Control System

This lab exercise adds an op-amp differentiator to the PI mode control system. The controlled process is a voltage source whose set point voltage is switched between 3 volts and 4 volts.

The transient and steady state response of a PID control system to a change of set point will be measured for several values of proportional gain, Kp, integral gain, Ki, and derivative gain, Kd. The responses will be compared to simulations. This lab experiment demonstrates the control system concepts of steady state error, response time, and settling time, overshoot, undershoot, and damping.

Parts and Equipment Required

Oscilloscope, DMM, Function Generator. Power Supply: ±9V to ±12V. Controller board PID-X1 (or build the circuit on a breadboard). Resistors: 10k, 15k, 22k, 1Meg, all ¼W, 5%. Capacitors: 10uF Non-polarized - ceramic.

Procedure Part 1: Proportional Mode

1. If you are not using the PID-X1 plug-in, layout and build the circuit below as carefully as possible.

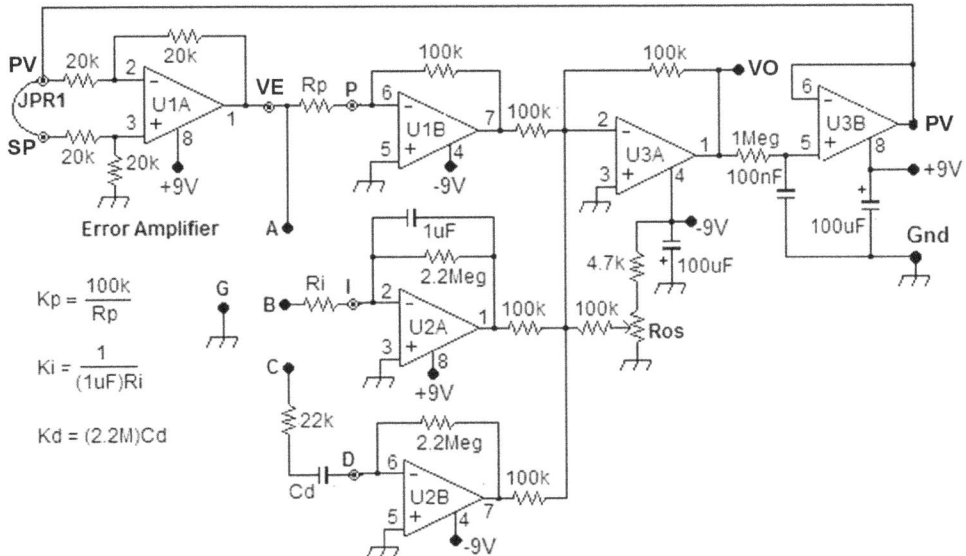

PID-X1 note: **Vpv** = **PV**, **Vsp** = **SP**, **Ve** = **VE**, **Vos** = **VO** when **Ve** = 0.

1b.   Controller board setup:

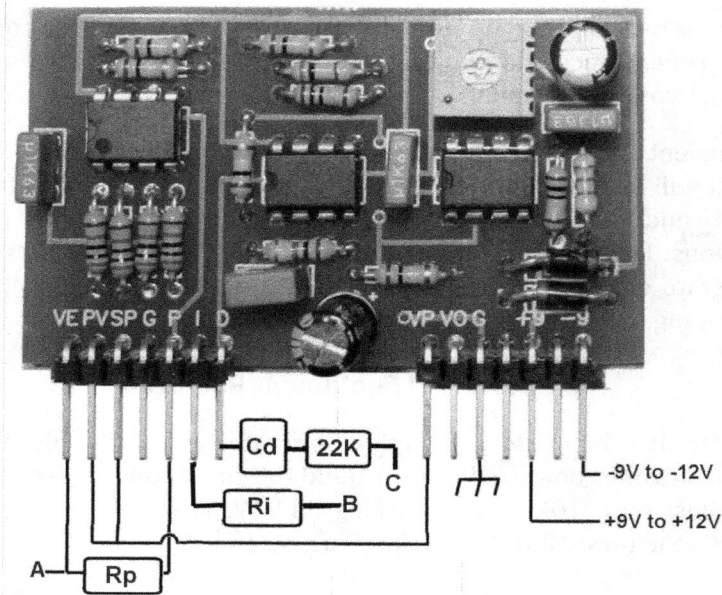

2.   **JPR1** should be connected. Connect a 1Meg resistor between points B
        and G (ground). Apply ±9 volts to the circuit. Check that the voltage
        VE is zero. It should be very close to zero since **PV = SP** because of
        the jumper wire, **JPR1** .

        Adjust the potentiometer, Ros, to set Vpv to exactly 3.0 volts. Use a
        DMM to make the measurement.

3.   Remove the jumper between **PV** and **SP**.

4.   Set the function generator to produce a 1 volt peak-to-peak, 1Hz,
        square wave with a 3.5 volt offset (so it goes between 3.0 volts and 4.0
        volts). Connect the function generator to **SP**.

5.   Connect the oscilloscope channel 1 to **Vsp** and Channel 2 to **Vpv**. Set
        vertical inputs to DC coupling and 200mV/Div. Set horizontal to
        100mS/Div and Trigger on channel 1.

6.   Record your results for Vsp below . If satisfactory, continue to part 2.

        Steady State _____volts   Rise Time _____ mS

        Settling Time _____mS

Procedure Part 2: Proportional Plus Integral Mode Check

1. Disconnect the 1Meg resistor between points **B** and **G** (ground). Connect point **B** to point **A**. Channel 1 should still be connected to **Vsp** and Channel 2 to **Vpv**. Set both vertical inputs to DC coupling and 200mV/Div. Time base is set to 100mS/Div and trigger is set to channel 1.

2. From your oscilloscope display, determine the steady state value, rise time, and settling time of **PV**. Record results for Rp = 15k and Ri = 10k below.

   Steady State _____volts   Rise Time _____ mS

   Settling Time _____mS

Procedure Part 3: PID Mode

1. Connect point **C** to point **A**. Channel 1 should still be connected to **SP** and Channel 2 to **PV**. Set both vertical inputs to DC coupling and 200mV/Div. Horizontal to 100mS/Div and trigger on channel 1.

2. Use: Rp = 15k, Ri = 10k, and Cd =10uF (non-polarized ceramic).

3. From your oscilloscope display, determine the steady state value, rise time, and settling time. Record results below.

   Steady State _____volts   Rise Time _____ mS

   Settling Time _____mS

4. Try to improve the system's response by repeating step 2 with other values of Ri and Cd.

Analysis

Write a report on this experiment using a word processor, spreadsheet, and simulation results. Use "cut and paste" techniques as appropriate.

Compare simulation results and experimental results. Document simulation and experimental results with graphs and calculations as appropriate. Include an abstract and conclusion.

Chapter 7: Real Systems

Practical circuits using feedback control system principles include power supplies, temperature controllers, position controllers, and speed controllers. The control concepts and circuits in the previous chapters can be adapted and scaled to a variety of applications. Seven applications are presented in this chapter: a variable voltage regulated power supply, a 2-point temperature controller, a proportional mode temperature controller, a proportional mode motor speed controller, PI motor speed controller, PID motor speed controller, and a magnetic Theremin.

Application 1: Variable Voltage Regulated Power Supply

This experiment demonstrates the application of an operational amplifier in a regulated power supply.   In this application the operational amplifier is a very high gain proportional controller.   A zener diode regulated voltage source and potentiometer are used to input the set point voltage to the non-inverting input of the op-amp.

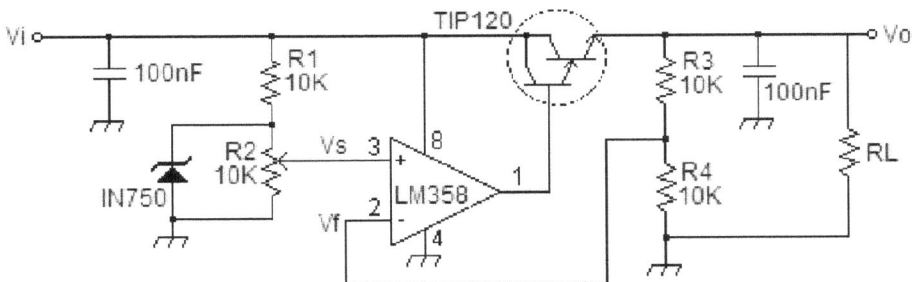

Feedback from the power supply's output voltage is applied to the inverting input of the op-amp.   The output of the op-amp is applied to the base of a darlington power transistor.   The transistor's emitter is the output of the power supply. Refer to and study the schematic diagram below.

Recall that the action of the op-amp in linear mode will be to make the voltage at its inverting input equal to the voltage at its non-inverting input (within a few millivolts). This means that Vf will equal the set point voltage, Vs. In this case Vf is one half of the supply's output voltage, Vo.

$$Vf = \frac{R4}{R3+R4}\, Vo = \frac{Vo}{2}, \quad \text{Therefore}: \ Vo = 2Vf = 2Vs.$$

With a 5 volt zener diode, Vs can vary from 0 volts to 5 volts and the output voltage, Vo, can vary from 0 volts to 10 volts.   You must use an LM358 to be able to vary the output voltage close to 0 volts, do not substitute. The input voltage, Vi, must be at least 2.5 volts higher than the maximum value of output voltage due to the op-amp's positive saturation voltage and the voltage drop across the transistor's emitters.

Parts and Equipment Required

Power supply: 10 to 15VDC, 1 amp.   Digital Multimeter (DMM).
Op-amp: LM358. Transistor: TIP120.   Zener diode: 1N750 or equiv.
Resistors: 3 - 10k, ¼ watt, 5%. 100Ω, 5 watt, 5%.
Capacitors: 2 – 100nF, 50V, 20%.

Procedure

1.   Connect the circuit on the previous page with RL = 1k. Connect the input, Vin, to a +12VDC power supply. Turn power on and verify that you can adjust the output voltage from less than 1 volt to more than 8 volts with R5.

2.   Set the output voltage, Vo, to 8.00 volts with R5. This will be the "no load value, Vo_{NL}.   Measure and record the Vf and Vs.

Vo_{NL}:_____   Vf:_____   Vs:_____

3.   Change value of RL to 100 ohms (5 watt resistor). Measure Vo. This will be the "no load value, Vo_{FL}. Measure and record the Vf and Vs.

Vo_{FL}:_____   Vf:_____   Vs:_____

4.   Measure the "line regulation" with RL = 1k ohms, for the regulator input voltage changing from 10.0 volts to 14.0 volts.

Vo_{10V}:_____   Vf:_____   Vs:_____

Vo_{14V}:_____   Vf:_____   Vs:_____

Analysis

1.   Calculate the power supply's load and line regulation (use equations below).   Indicate at least two ways to improve load regulation and line regulation. Write a report on the results of your experiment.

$$\% Load\,Reg = \frac{Vo_{NL} - Vo_{FL}}{Vo_{FL}} 100\% \qquad \% Line\,Reg = \frac{Vo_{14V} - Vo_{10V}}{Vo_{10V}} 100\%$$

70

Application 2: Two Point On/Off Temperature Controller

An on/off temperature controller connects power to the heater when the temperature is below the set point and disconnects the power when the temperature is above the set point. Practical on/off controllers have two set points, a low temperature limit and a high temperature limit. This reduces the number of times the system cycles on and off.

This lab exercise uses a Schmitt trigger for the controller. Its output saturates at 0 volts and at about 8 volts with a 9 volt power supply. The output is applied to an enhancement mode power mosfet transistor, Q1. Q1 acts as a switch that controls the power applied to a heater. The heater is a 10 ohm, 5 watt resistor. A temperature sensor IC, the LM35, senses the temperature.

Parts and Equipment Required

Oscilloscope / Data Logger, DMM. Power Supply: 0 to 6V and +9 volts.
IC: LM358, Transistor: IRF510 or equiv.
Resistors: three 10k, 15k, 91k, 150k, ¼W, 5%.
Capacitors: 10µF, 10nF.
Oven parts: 10 ohm, 5 watt resistor, LM35 IC, in small box.
Refer to appendix 1 for oven information. Alternately, the 10Ω heater resistor can be plugged into a breadboard with the LM35 next to it.

Procedure

1. Connect the circuit below. Use separate ground leads for the 9 volt power supply and the 0 to 6 volt supply. Connect the ground lead of the 0 to 6 volt supply close to Q1's source. Do not connect the 0 to 6 volt power supply to the oven yet.

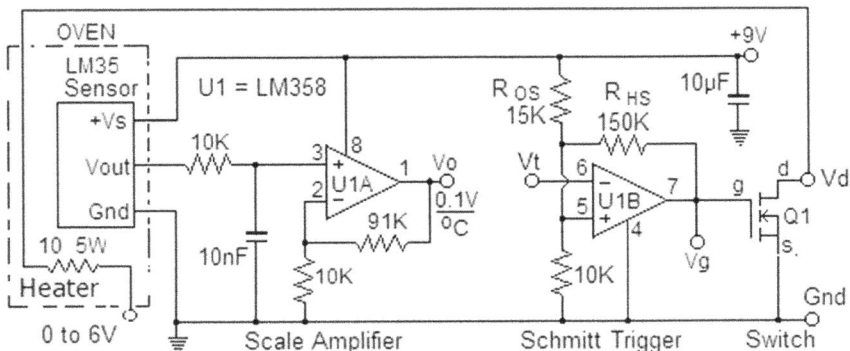

2. Connect a DMM to terminal **Vo** to monitor the temperature. It should read room temperature (about 2.0 volts if the room is at 20° Celsius).

    The thermal characteristics and connection diagram for the LM35 is shown on the right. **Vout** = 10mV per degree Celsius.

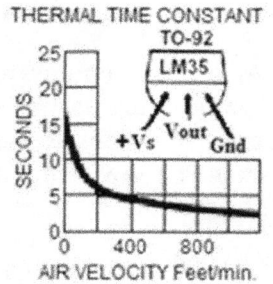

THERMAL TIME CONSTANT

3. Check the operation of the Schmitt trigger control circuit using the following procedure.

    Connect a function generator to terminal **Vt** set to produce a 5 volt peak to peak, 20 Hertz triangle wave, offset by 2.5 volts.

    Connect channel 1 of the oscilloscope to terminal **Vg** and channel 2 to terminal Vt. Set the trigger to channel 1.

    Set both channels to 1 volt per division and DC coupling. Set the zero reference to the bottom of the screen. Set the time base to 10mS per division.

    Oscilloscope display should be similar to the one on the right.

    The objective is to determine the voltage values of the triangle wave that trigger the Schmitt trigger circuit.

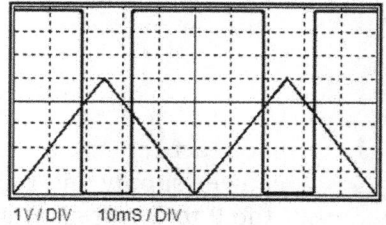

1V / DIV   10mS / DIV

    The display shows that **Vg** goes low when the triangle wave reaches about 3.7 volts on the positive slope and goes high when the triangle wave goes below about 3.4 volts. Your set points may be slightly different.

4. The "process variable" of this control system is the oven temperature, which is represented by the voltage Vt, which is applied to the trigger input of the controller. **Vt** = 0.1 volts per degree Celsius. 3.4 volts = 34 degrees and 3.7 volts = 37 degrees.

    Increasing the value of R_{HS} decreases the separation between the high trigger voltage and the low trigger voltage (between the high set point and low set point). Increasing the value of R_{OS} raises both the high and low trigger points.

    Disconnect the function generator from terminal **Vt.**

72

5.   Connect terminal **Vo** to terminal **Vt**. Prepare channel 1 of your data logger to log the voltage at terminal **Vg** and channel 2 to log the voltage at terminal **Vt**. **Vg** will be close to zero when the oven is off and close to 8 volts when the oven is on.

6.   Connect the DMM to Vt to monitor the oven temperature.

7.   Set the data logger sample rate to one sample every 2 seconds and the number of samples to 300. This may need to be adjusted depending on your oven's/resistor's response.

8.   Set the 0 to 6 volt supply to 4.0 volts. Start the data logger and immediately connect the power supply to the oven.   You should be observing the temperature rise. When the temperature reaches the upper trip point you should observe that channel 1 voltage drops to zero. If this does not happen, check your connections and circuits. Log about two to three cycles of temperature data.

.   Turn the power to your experiment off. Open your acquired data in a spreadsheet.   Graph the data, labeling the y axis in degrees Celsius, and the horizontal axis in seconds.

Analysis

1.   From your logged data determine:

        a)   Temperature cycle time.
        b)   Controller's low set point and high set point.
        c)   Temperature over-shoot and under-shoot.

2.   From your logged data determine:

        a)   Oven's heating transfer function in degrees C per second.
        b)   Oven's cooling transfer function in degrees C per second.

Simulation of On/Off Control

The oven is simulated by using the RC time constant of R1 and C1 to model its heating and cooling time. Remember to right click on the mosfets to select the IRF510. **Vpv** is connected to the inverting input of the Schmitt trigger. **Vsp** is on the non-inverting input. **Vsp** alternates between the upper and lower set points.

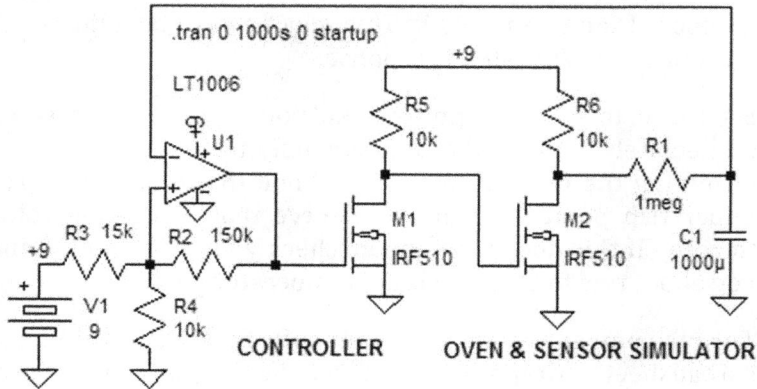

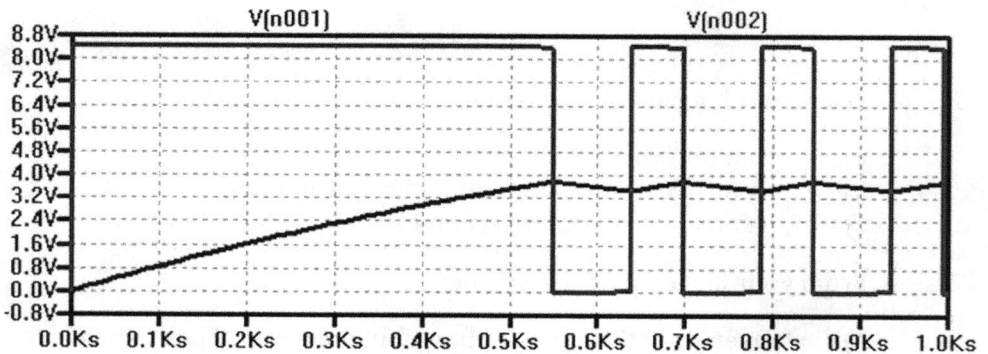

Simulation results above show that the upper set point is first reached in about 550 seconds. The oven cools to the lower set point in about 90 seconds and reheats to the upper set point in about 60 seconds. The graph can be expanded as shown below. Click on the vertical and horizontal axis to change the range.

74

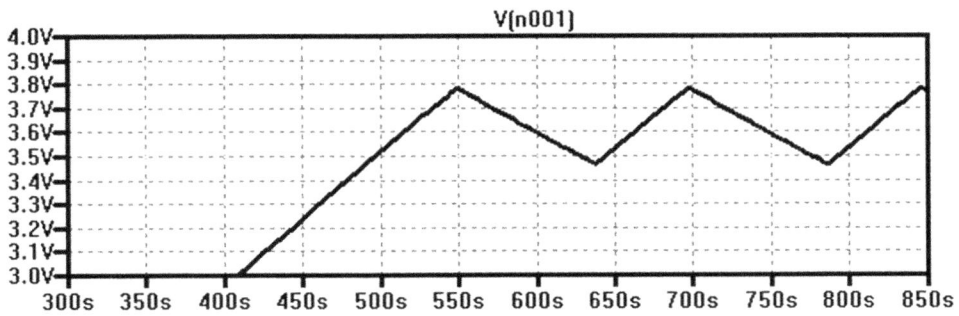

Heating transfer function: $T_H = \dfrac{38^0 - 35^0}{60S} = \dfrac{0.05^0}{S}$.

Cooling transfer function: $T_H = \dfrac{35^0 - 38^0}{90S} = -\dfrac{0.033^0}{S}$

Application 3: Proportional Mode Temperature Control

This lab exercise demonstrates the proportional mode feedback control system as used to control the temperature of a small oven (or resistor). A temperature sensor IC measures the oven temperature. The heater is a 10 ohm 5 watt resistor. An amplifier scales the output of the temperature sensor to 0.1 volt per degree Celsius. A differential amplifier is used as the error detector. See the block diagram below.

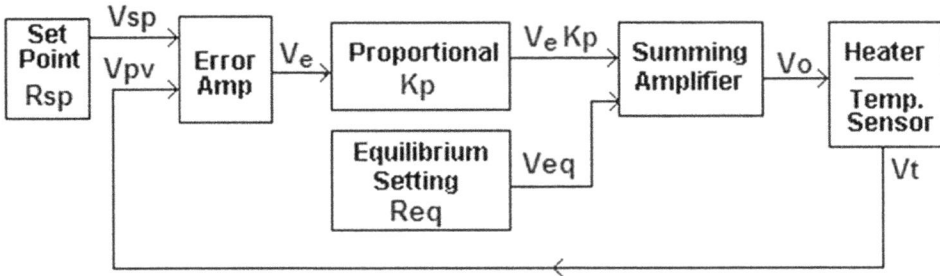

The response of the control system to a change of set point will be measured for several values of amplifier gain. In Part 1 the response of the oven will be measured. In part 2 the proportional only mode of temperature control will be implemented. This lab experiment demonstrates the control system concepts of proportional band, proportional error, response time, and settling time.

Analysis is simplified by setting the gains of the error amplifier and summing amplifier to one. The system's gain is set entirely by the proportional controller. The transfer function of the sensor circuit is 0.1 volts per degree Celsius. The transfer function of the oven depends on physical properties of the oven and the voltage applied to the heater. It has units of degrees per second.

Parts and Equipment Required

Oscilloscope / Data Logger, DMM.
Power Supplies: 0 to 6V, 2 amps; ±9V to ±12V, 100mA.
PID-X1 controller board (or build circuit on a breadboard)
Transistor: TIP120 or equiv.
Resistors: 220, 820, 2-10k, 20K, 91k, 100k, ¼W, 5%.
Potentiometers: 1k trimmer.
Capacitor: 10nF.
Oven: Refer to appendix 1 (10 ohm, 5 watt resistor and LM35 temperature sensor).

Procedure Part 1: Oven Response

1. Connect the temperature sensor circuit shown on the right.

    Turn on the +9 volt supply (can be 9 to 12 volts). <u>Do not</u> connect the 0 to 6 volt power supply to terminal **Vp** yet. Connect a DMM to terminal **Vt** to read the temperature.

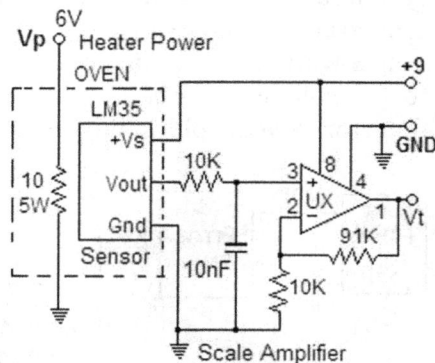

2. Measure the voltage at **Vt**. It should represent the room temperature (2.0 volts if the room temperature is 20 degrees Celsius).

3. Use the DMM to monitor the voltage **Vt**.

4. <u>Check that the oven lid is closed and oven temperature is below 30⁰ Celsius.</u> Set the 0 to 6 volt heater supply to 2.0 volts. Connect the heater power supply to terminal **Vp**.

    You should observe the voltage **Vt** increasing. Wait until the voltage stops changing or is changing very slowly (oven reaches equilibrium when the temperature changes by less than about 0.1 degree in a time interval of about 20 seconds).

    Record the temperature. Temperature in degrees Celsius is equal to ten times the voltage, **Vt**.

    $T_{EQ=2.0V}$ _____

5. Increase **Vp** to 3.0 volts. Wait for the oven to reach equilibrium temperature. Record.

    $T_{EQ=3.0V}$ _____

6. Estimate the value of **Vp** for an oven temperature of 36⁰ C from the measurements of steps 4 and 5 above. Adjust **Vp** to your estimated value and wait for the oven to reach equilibrium.

    Readjust **Vp** if necessary until the equilibrium temperature is within about 2.0 degrees of 36 degrees (34 t0 38). Record the value of **Vp$_{EQ}$** and the equilibrium temperature below:

    T_{EQ} _____ Vp_{EQ} _____

Changes in room temperature, air circulation, and location of the oven may affect this value of Vp_{EQ}. Keep the oven in the same location during the rest of this experiment. Be sure that the lid is on the oven when taking data.

Procedure Part 2: Proportional Control

1. The control circuit schematic diagram is given below. Layout and build the circuit as carefully as possible. Keep the number of wires to the absolute minimum. Observe the numbers on the ICs. For example, U1A and U1B is in the same package.

PID-X1 note: **Vpv = PV, Vsp = SP, Ve = VE, Vos = VO** when **Ve** = 0.

If you are using the PID board, connect a 1 megohm resistor from pin **Vi** to ground (this prevents the integrating capacitor from charging).

2. Be sure that the jumper wires, **J1**and **J2** are connected as shown. Study the schematic diagram to understand the function of these jumpers as described below.

J1: Tests the controller in "open loop" mode. **J1** will be removed and **Vpv** will be connected to **Vt**, the output of the temperature sensor circuit, for closed loop operation.

J2: Pulling out **J2** will cause the set point to increase by about 5 degrees (0.5V).

Steps 3, 4, and 5 perform an initial setup and check of the control circuit. If the response is not what is expected, check the power supply connections, check the input and output voltages of the op-amps, and check the wiring carefully.

3.   Apply plus and minus 9 to 12 volts to the controller and plus 6 volts to the heater circuit (TIP120 collector). Set the 0 to 6 volt supply to 6.0 volts. The temperature sensor circuit is not connected to the control circuit yet. Check that the voltage **Ve** is zero. Set **Vp** to zero with the potentiometer Req.

If **Vp** or **Ve** are not zero (less than 0.1 volts), locate the problem by checking voltages at the inputs and outputs of each stage.

Adjust the potentiometer, Ros, to set **Vp** at the emitter of the transistor to the value of **Vp$_{EQ}$** which you measured in part 1, step 6.

4.   Set the voltage **Vsp** with potentiometer Rsp to the voltage that represents the oven's equilibrium temperature (T$_{EQ}$ from step 6 of part 1). This is the set point voltage. For example, if T$_{EQ}$ is 37 degrees, set **Vsp** to 3.7 volts.

5.   Double check that **Vsp** is correct (set point voltage) and that **Ve** is zero. Turn off the 0 to 6 volt power supply

Data Logger Setup

1.   Remove jumper **J1** and connect **Vpv** to **Vt** (control circuit input to temperature sensor circuit output). Connect data logger channel 1 to measure **Vpv** (oven temperature) and connect data logger channel 2 to measure **Vsp** (set point temperature).

Channel 1 will log the oven temperature as it approaches the set point. If more channels are available, error voltage **Ve** and oven voltage **Vp** could be logged.

2.   Set the data logger to take one sample every two seconds. Set the number of samples to 400 (logging can be stopped when sufficient data is logged).

3.   Connect the DMM to **Vt** to monitor the temperature. The oven temperature should be below 30 degrees before starting the data logger. If it is above 30 degrees take the lid off the oven and cool it with a paper fan.

4.   When ready, with the lid on the oven, apply power to the oven by turning on the 0 to 6 volt power supply and start the data logger.

When the oven temperature stops changing (reaches equilibrium) for about 20 seconds, pull out jumper **J2**. This will cause the oven temperature to increase to the new set point. When the temperature stops changing, turn off the data logger and the oven power supply.

5.　　Cool the oven to a temperature below 30 degrees. Reconnect jumper **J2**. Change Rp to 20k ohms. Repeat steps 2, 3 and 4.

Note:　The oven temperature should reach the first set point accurately. When the jumper wire is pulled the set point is increased and the temperature will increase, but it will not reach the second set point. Step 5 increases the proportional gain so that the temperature will get closer to the second set point, but it will still not reach it.

You may need to adjust your data logger timing to better match your oven's response.

You may need to check your oven equilibrium calibration and controller setup if the temperature does not reach the first set point accurately.

Analysis

Write a report on this experiment. Use a word processor and cut and paste operations to transfer all relevant data and graphs to your word processor document. Include the following:

1.　　Introduction and objectives.

2.　　Data and graphs for each part.

3.　　Analysis of data and graphs.

a)　　System response from room temperature to first set point.
b)　　System response from first set point to second set point.
c)　　Rise times, settling times, overshoot, undershoot, oscillation.
d)　　Given that the proportional range of the heater voltage is about 0 volts to 5 volts, determine the proportional band of the control system for the error voltage, Ve, for each value of Rp.

4.　　Summary and conclusion.

Motor Speed Control

The next three applications should be done sequentially, starting with application 4. All three labs require a small DC motor with an optical interrupter that produces 4 pulses per rotation. All three applications use the PID plug-in board described in Appendix 2. If the PID board is not used, the circuit may be built on a breadboard. Refer to Appendix 2 for the circuit diagram and parts information. Parts required in addition to the PID plug-in board are listed in each application. A suggested motor-tachometer assembly is described in appendix 1.

Application 4: Proportional Mode DC Motor Control

Tachometer

An "optical interrupter module" and a "mono-stable multi-vibrator" (one-shot) are used to measure the rotational speed of a small DC motor. The motor's shaft has a wheel attached which has holes or slots in it through which the light from an LED in the interrupter module can pass. Resulting light pulses are converted to voltage pulses by a photo-transistor. These pulses trigger the one-shot.

Each trigger pulse causes the one-shot to produce a constant width pulse whose width is independent the trigger frequency. Since the pulse width is constant, the duty cycle at the output of the LM555 increases with frequency. This produces a voltage at the output of the low-pass filter/integrator which is proportional to the motor's rotational speed (rpm). The output pulse width is adjusted to produce a transfer function of 1 volt per 1000 rpm of the motor. Refer to the block diagram below showing the motor and tachometer circuit connected to a PID control system.

A block diagram of the motor-tachometer circuit is shown below. Potentiometer **Rpw** sets the output pulse width of the one-shot. It is is adjusted so that the voltage, **Vpv**, is equal to 1 volt per 1000 rpm. Refer to the table on the next page.

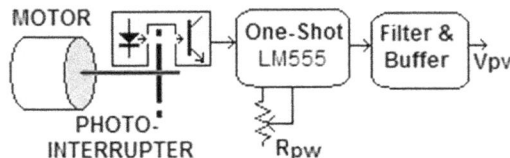

RPM	Pulses/Min.	Pulses/Sec.	Period -mSec.	Duty Cycle	**Vpv**
2000	8000	133	7.5	.29	2.0
3000	12000	200	5	.43	3.0
4000	16000	267	3.75	.57	4.0
5000	20000	333	3	.71	5.0

The table above assumes that the photo-interrupter generates 4 pulses per revolution of the motor.   The one-shot's trigger pulse width must be shorter than its output pulse width.   The output voltage **Vpv** also depends on the one-shot's output voltage, which is about 7 volts when an 8 volt power supply is used. The values in the table assume a one-shot output voltage of 7.0 volts.

Parts and Equipment Required

Dual-trace Oscilloscope, Digital Multi-meter
PID-X1 controller board (or build circuit on a breadboard)
Power Supply: 6 volt, 1 amp and ±12 volts, 100mA.
ICs: LM555, LM358, LM78L08. Transistor: TIP120 or equiv.
Resistors: 4.7K, 10k, 47k, 100k, 1/4W, 5%, 2-10K trim pots.
Capacitors: 2 - 10nF, 470nF, 2-10μF electrolytic.
DC Motor/Photo-interrupter.   Refer to the motor set in Appendix 1.

Procedure Part 1: Tachometer Calibration

1. Connect the tachometer circuit as shown below.

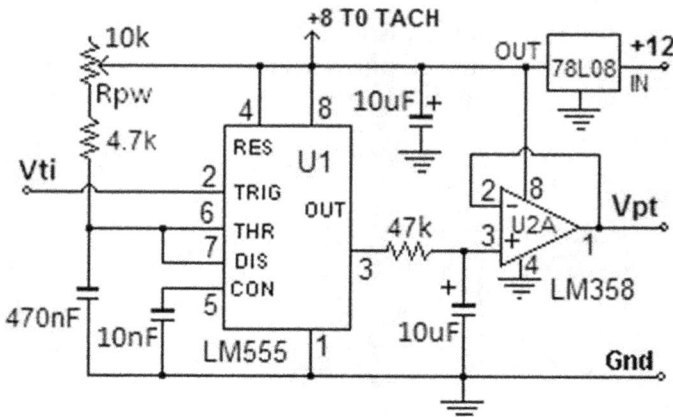

84

Connect a function generator to **Vti** set to produce a 200 Hertz, 8 volt peak-to-peak square wave, with a 20% duty cycle, and with a positive offset of 4 volts (so it goes from 0 to 8 volts). See the note below step

3. Connect oscilloscope channel 1 to pin 2 of the LM555. Connect Channel 2 to pin 3 of the LM555. Trigger on channel 1, on falling edge.

    Note: If you can't vary the duty cycle of the function generator, connect the high-pass filter circuit on the right between the generator and **Vti**. The diode clamps the positive pulse. This produces a sharp negative pulse with a   short exponential rise time.

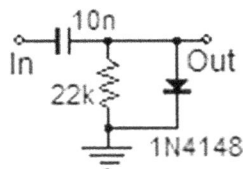

    The waveforms on channels 1 and 2 should be similar to the simulated waveforms on the right. Channel 1 shows the trigger pulse applied to the trigger input.

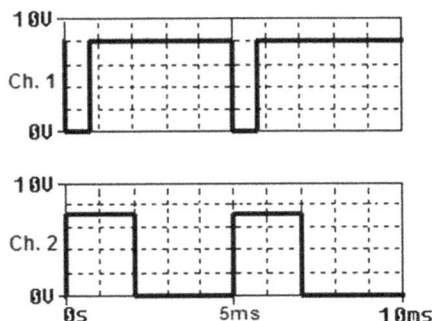

    Channel 2 shows the output pulse of the LM555. You can vary the pulse width with the   pot, **Rpw**.

4. Connect a DMM to read the DC voltage, **Vpt**, at pin 1 of U2. Adjust the potentiometer, **Rpw**, to produce exactly 3.0 volts on the DMM.

5. Measure and record the pulse amplitude in volts and the pulse width in milli-seconds of the channel 2 waveform.

    Amplitude (200 Hz): _____   Pulse width (200 Hz): _____

6. Change the function generator frequency to 267 Hertz. Verify that the DMM reads a voltage close to 4.0 volts (about ±0.1 volts).

    This completes the calibration of the tachometer. Disconnect the function generator.

Procedure Part 2: Motor Characteristics

The diagram below shows the motor and photo-interrupter circuit. The motor is driven by the darlington transistor, TIP120, whose emitter voltage (and motor voltage) is controlled by the voltage applied to its base.

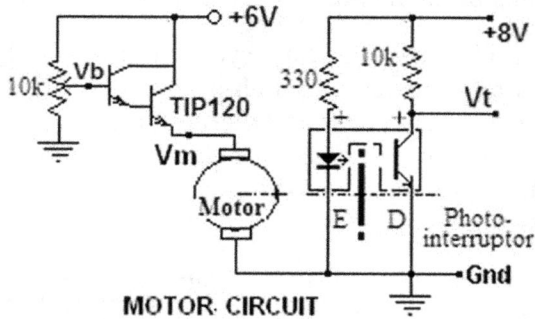

MOTOR CIRCUIT

Note that the output of the photo-interrupter is at the terminal **Vt**. **Vt** will be connected to the input, **Vti**, of the tachometer circuit. The 8 volt power supply for the photo-interrupter circuit is from the 8 volt regulator of the one-shot circuit.

1. Connect a the motor circuit shown above but do not connect the TIP120 emitter to the motor yet. Disconnect the function generator. Connect **Vti** of the tachometer circuit to **Vt** of the motor-generator circuit. Also connect **Vs** to the output of the 8 volt regulator. Be sure to run a separate ground wire from the 6 volt power supply directly to the motor ground.

2. Set **Vb** to zero volts with Rs. Connect the emitter of the TIP120 to the motor (**Vm**).

3. Connect the DMM to **Vpv** to measure the motor speed (1 lt/1000rpm). Adjust the potentiometer, Rs, to set the motor speed to 2000 RPM (Vr = 2.0 volts).

   Measure the resulting voltages Vb and Vm (by moving the DMM probe). Repeat for 3000 RPM and 4000 RPM. Record below.

Speed - RPM	Vb	Vm
2000		
3000		
4000		

86

Note: The motor speed will fluctuate some, but if the fluctuation is greater than 200 rpm, check the 6 volt power supply connections. Be sure to use a separate ground lead from the 6 volt power supply directly to the motor ground.   We will see later that the feedback control circuit will reduce the motor speed fluctuation.

Part 3: Proportional Control

The motor-tachometer circuit is connected to the proportional controller as shown in the block diagram on the next page.   This system's response is determined by its proportional gain, Kp, and the transfer function of the motor-tachometer circuit.

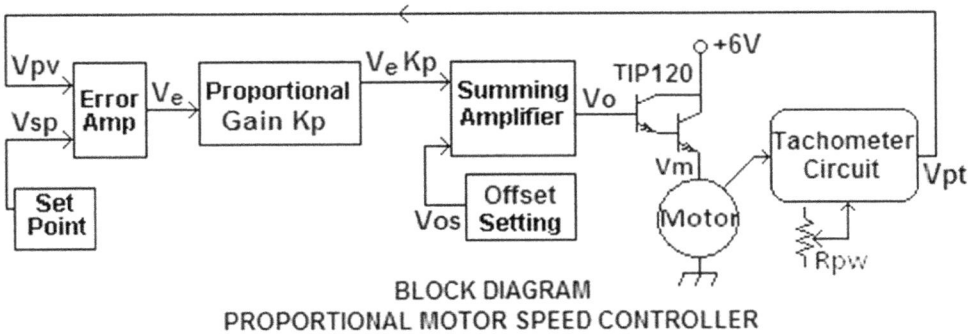

BLOCK DIAGRAM
PROPORTIONAL MOTOR SPEED CONTROLLER

The steady state transfer function of the motor-tachometer circuit can be approximated as the product of the motor's transfer function, (rpm/**Vo**), and the tachometer's transfer function, {**Vpt**/rpm). This product gives the gain of the motor-tachometer circuit as (**Vpt/Vo**).

The motor-tachometer circuit transfer function also includes a transient time dependent component. The motor is able to change its speed by 1000 rpm in less than 0.1 seconds. However, the response time of the filter in the tachometer circuit is significantly longer so that the transient response of the motor-tachometer circuit can be approximated as the response of the tachometer circuit's filter.

1.   Connect the proportional mode control circuit to the motor-tachometer circuit as shown on the next page, including jumper **J1**.

        If you are using the PID board, connect a 1 megohm resistor from pin **I** to ground   (this prevents the integrating capacitor from charging). Note that the controller's output, **Vo**, is not yet connected to the motor circuit input, **Vb**.

2. Measure the error voltage, **Ve**. If it is not close to 0.0 volts, check your circuit connections.   Connect a DMM to measure the controller's output voltage, **Vo**. Set **Vo** to 0.0 volts with the potentiometer, **Ros**.

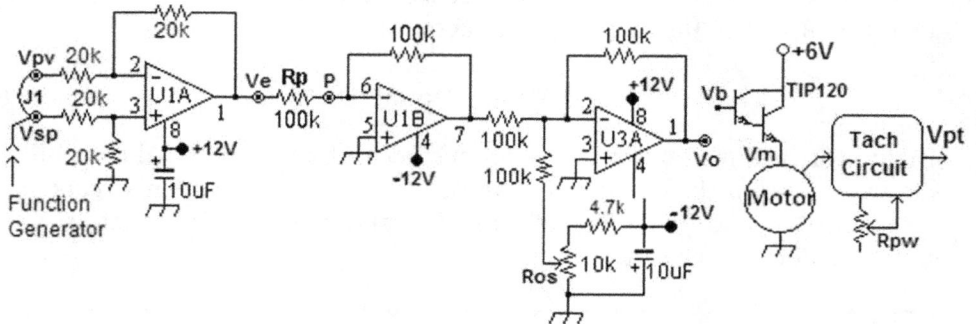

PID-X1 note: **Vpv = PV, Vsp = SP, Ve = VE, Vos = VO** when **Ve** = 0.

3. Connect the controller's output, **Vo**, to the motor circuit's input, **Vb**. Connect the DMM to the tachometer's output to measure the voltage, **Vpt**. Adjust the potentiometer, **Ros**, to set **Vpt** to 2.50 volts. The motor should now be running at 2500 rpm. The speed may fluctuate some. We will see that this fluctuation will be reduced when feedback is applied.

4. Set the function generator to produce a 0.2Hz, 1 volt peak-to-peak square wave with a plus 3 volt offset. Connect the function generator and channel 1 of the oscilloscope to **Vsp**. Connect channel 2 of the oscilloscope to **Vpv**. Do not remove the jumper **J1** yet.

5. Set channel 1 and channel 2 of the oscilloscope to DC input and 0.5V/DIV. Set the time base to 0.5S/DIV and trigger to channel 1.

Set the vertical position   controls of both channels so that 0.0   volts is at the bottom of the display   and 4.0 volts is at the top.

At this point both channels should have   identical waveforms. Your display should be similar to the one shown on the right.

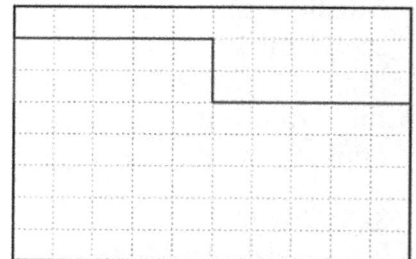

VERTICAL: 0.5V/DIV HORIZONTAL: 0.5S/DIV

6. Remove the jumper, **J1**. Connect **Vpt** to **Vpv**. The motor speed should be changing between 2500 rpm and 3000 rpm.

88

The oscilloscope display should be similar to that shown on the right.

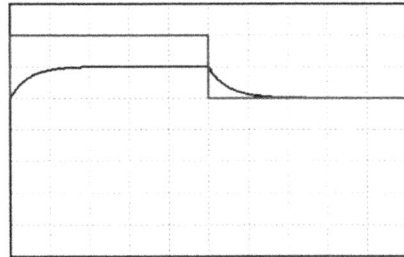

Note: Your response may vary depending on the type of motor used. Be sure that **Vpv** settles on 2.5 volts when **Vsp** = 2.5 volts.

Vpv will look "fuzzy"   due to the ripple from the tachometer filter. Use the average value of **Vpv** for your data.

Record the following data for proportional gain equal to 1:

Vpv rise time: _____   Vpv fall time: _____

Vpv high settling voltage: _____

Vpv low settling voltage: _____

7.   Repeat step 6 for proportional gain equal to 10 (change the value of Rp to 10k).

Record the following data for proportional gain equal to 1:

Vpv rise time: _____   Vpv fall time: _____

Vpv high settling voltage: _____

Vpv low settling voltage: _____

8.   Optional: Experiment with higher proportional gains. What value of gain produces the best response in terms of rise time, settling time, and proportional error?

Analysis

Write a report on this experiment.   Use a word processor and "cut and paste" to transfer all relevant information. Include the following:

1.   Introduction and objectives.

2.   Data and graphs for each part.

3.   Analysis of data and graphs.

4.   Summary and conclusion.

Application 5: Proportional Plus Integral Motor Speed Control

This application is a continuation of application 5. An integrator is added to the proportional controller as shown in the block diagram below.

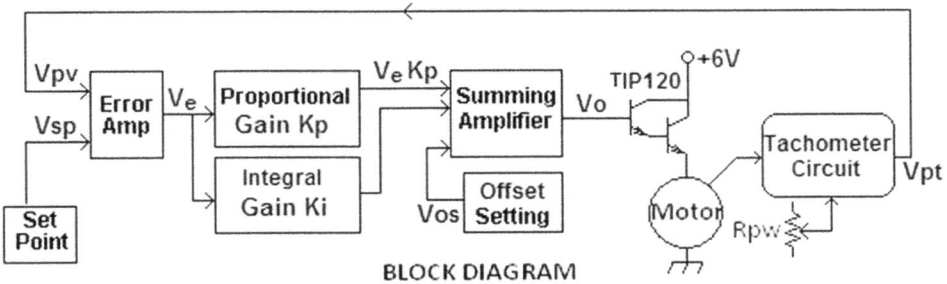

BLOCK DIAGRAM

A review of the PID control components and operational amplifier implementation is presented below.

Proportional

The proportional controller's output is:

$$Vop = -K_P \cdot V_e. \qquad K_P = \frac{Rf}{Rp}$$

The integral controller's output is:

Integral

$$Voi = \frac{-1}{Ri \cdot C}\int_0^t V_e d\tau + Vi. \qquad Vi = Voi \, at \, t = 0.$$

If V_e is constant, $Voi = -Ki \cdot V_e \cdot \Delta t \ \Rightarrow\ Ki = \frac{1}{Ri \cdot C}.$

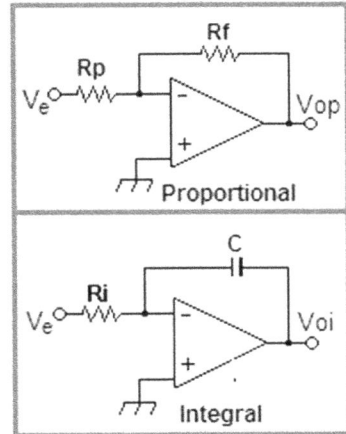

Controller's response, Vo, is given by the sum of the responses:
$$Vo = Vop + Voi.$$

Parts and Equipment Required

(Most of the equipment and parts are the same as in application 4)

Procedure

Note: Check the proportional mode control circuit of application 4. Reset the proportional gain to 1 and make sure that the motor's equilibrium speed is still 2500 rpm.

1. Add the integrator circuit (U2A) as shown in the diagram below, but <u>do not</u> connect the function generator or turn on the power supplies. Jumper **J1** should be connected as shown. Rp = 100k, Ri = 470k. (Kp = 1, Ki = 2)

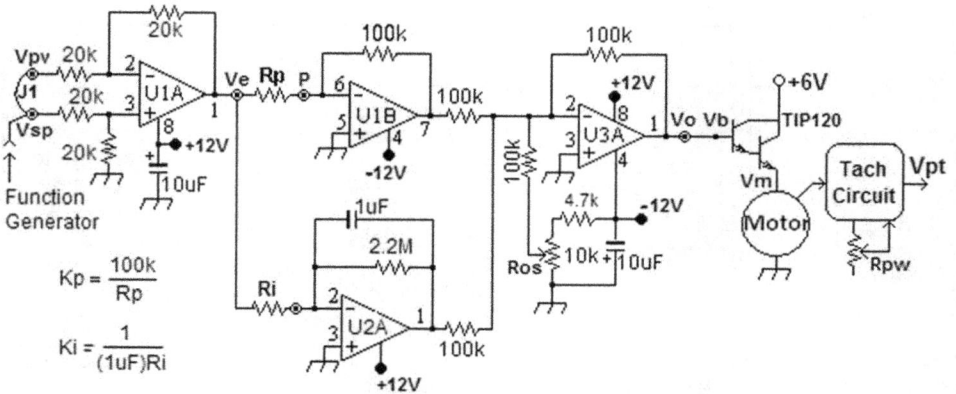

PID-X1 note: **Vpv = PV, Vsp = SP, Ve = VE, Vos = VO** when **Ve** = 0.

2. Turn on the power supplies. Measure the voltage, **Vpt,** at the output of the tachometer circuit with a DMM.

   Vpt: _____   RPM: _____

   Fine adjust equilibrium potentiometer, **Req**, so that **Vpt** is 2.5V ±0.1V, if necessary (Motor speed should be close to 2500 rpm).

3. Set the function generator to produce a 0.2Hz, 1 volt peak-to-peak square wave with a plus 3 volt offset. Connect the function generator and channel 1 of the oscilloscope to **Vsp**. Connect channel 2 of the oscilloscope to **Vpv**. Do not remove the jumper **J1** yet.

4. Set channel 1 and channel 2 of the oscilloscope to DC input and 0.5V/DIV. Set the time base to 0.5S/DIV and trigger to channel 1.

Set the vertical position controls of both channels so that 0.0 volts is at the bottom of the display.

At this point both channels should have identical waveforms. Your display should be similar to the one shown above on the right.

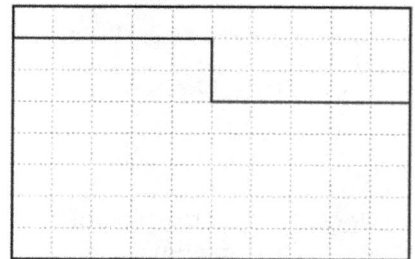

VERTICAL: 0.5V/DIV   HORIZONTAL: 0.5S/DIV

5.   Remove the jumper, **J1**. Connect **Vpt** to **Vpv**. The motor speed should now be changing between 2500 rpm and 3xxx rpm.

Note: Your response should be similar to that shown on the right. It may vary depending component tolerances and on the type of motor used.   You may need to adjust the value of **Ri** for the best response without oscillation.

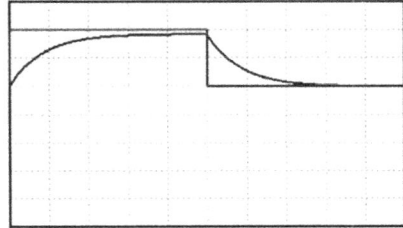

VERTICAL: 0.5V/DIV HORIZONTAL: 0.5S/DIV

Be sure that **Vpv** settles on 2.5 volts when **Vsp** = 2.5 volts.

Vpv will look "fuzzy" due to the ripple from the tachometer filter. Use the average value of **Vpv** for your data.

6.   Record the following data for proportional gain of 1 and integral gain of 2 (or other value):

Vpv rise time: _____   Vpv fall time: _____

Vpv high settling voltage: _____

Vpv low settling voltage: _____

7.   Repeat step 6 for proportional gain of 10. (change the value of Rp to 10k) and integral gain of 20 (change the value of Ri to 47k or one tenth of the value used in step 6). You may need to experiment with the integral gain to get a good response without oscillation.

Record the following data for proportional gain of 10 and integral gain of 20 (or other value):

Vpv rise time: _____   Vpv fall time: _____

Vpv high settling voltage: _____

Vpv low settling voltage: _____

8.   Optional: Experiment with other values of proportional gains. What value of gain produces the best response in terms of rise time, settling time, and proportional error?

Analysis

Write a report on this experiment. Use a word processor and "cut and paste" to transfer all relevant information. Include the following:

1. Introduction and objectives.

2. Data and graphs for each part.

3. Analysis of data and graphs.

4. Summary and conclusion.

Application 6: PID Control

A block diagram of the PID control system is presented below.

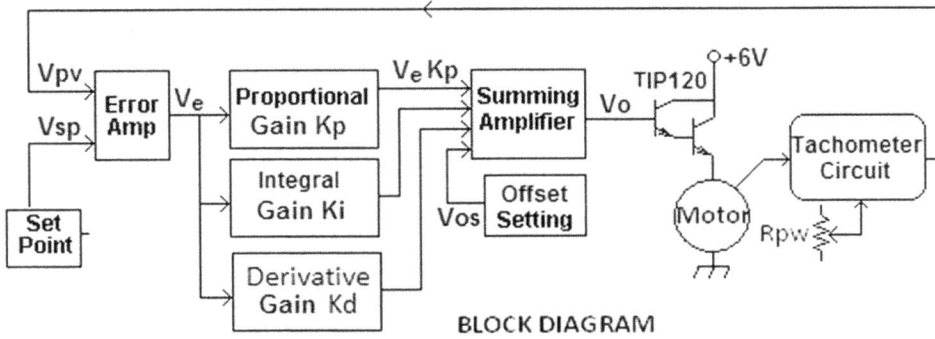

BLOCK DIAGRAM

Review of the PID control components:

The proportional controller's output is:

$$Vop=-K_P \cdot V_e. \qquad K_P=\frac{Rf}{Rp}$$

The integral controller's output is:

$$Voi=Ki\int_0^t V_e \, d\tau+V_{t=0}. \qquad Ki=\frac{1}{Ri \cdot C}.$$

if V_e is constant, $Voi=-Ki \cdot V_e \cdot \Delta t + V_{t=0}$.

The derivative controller's output is:

$$Vod=-Kd\frac{dV_e}{dt}. \qquad Kd=Rd \cdot C$$

If $\dfrac{dV_e}{dt}$ is constant for an interval Δt, then

$$Vod=-K\Delta d\frac{\Delta V_e}{\Delta t}$$

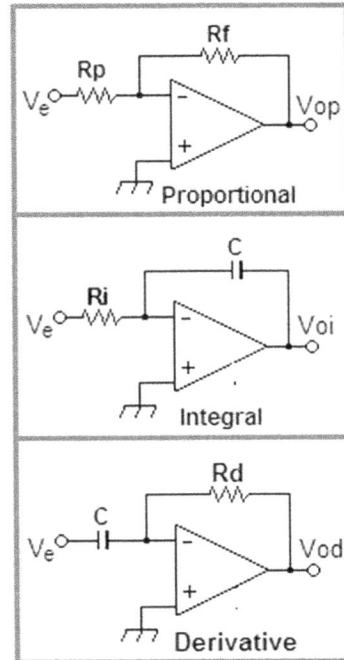

Controller's response, Vo, is given by the sum of the responses:

$$Vo=Vop+Voi+Vod \cdot$$

Parts and Equipment Required

(Most of the equipment and parts are the same as in application 4)

Procedure

Note: Check the proportional plus integral mode control circuit of application 4. The proportional gain should still be 10 and the integral gain should be 20 or the value that results in the best response. Rp =10k, Ri = 47k or the value used in application 4.

1. Add the differentiator circuit (U2B) as shown in the diagram below, but <u>do not</u> connect the function generator or turn on the power supplies. Jumper **J1** should be connected as shown. **Vo** should be connected to **Vb**. Cd =10uF (Differential gain = 22).

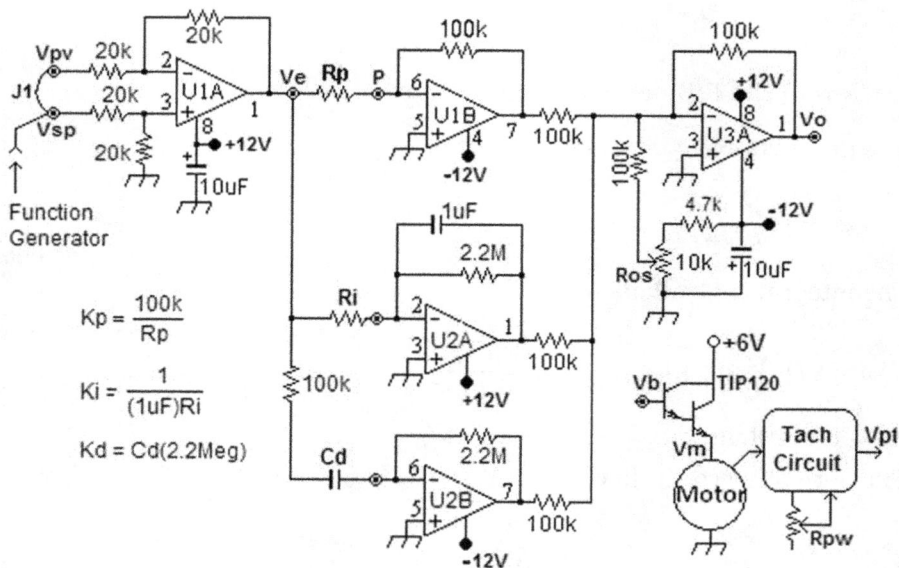

$$Kp = \frac{100k}{Rp}$$

$$Ki = \frac{1}{(1uF)Ri}$$

$$Kd = Cd(2.2Meg)$$

PID-X1 note: **Vpv = PV, Vsp = SP, Ve = VE, Vos = VO** when **Ve = 0.**

2. Turn on the power supplies. Measure the voltage, **Vpt,** at the output of the tach circuit with a DMM.

Vpt: _____   RPM: _____

Fine adjust offset potentiometer, **Ros**, so that **Vpt** is 2.5V ±0.1V, if necessary (Motor speed should be close to 2500 rpm).

3. Set the function generator to produce a 0.2Hz, 1 volt peak-to-peak square wave with a plus 3 volt offset. Connect the function generator and channel 1 of the oscilloscope to **Vsp**. Connect channel 2 of the oscilloscope to **Vpv**. Do not remove the jumper **J1** yet.

4. Set channel 1 and channel 2 of the oscilloscope to DC input and 0.5V/DIV. Set the time base to 0.5S/DIV and trigger to channel 1. Set the vertical position controls of both channels so that 0.0 volts is at the bottom of the display.

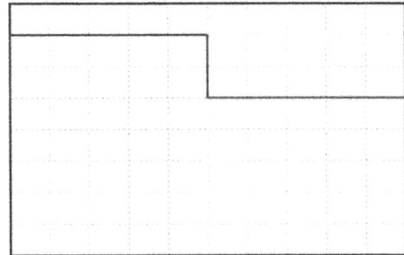

VERTICAL: 0.5V/DIV   HORIZONTAL: 0.5S/DIV

At this point both channels should have identical waveforms. Your display should be similar to the one shown above.

5. Remove the jumper, J1. Connect **Vpt** to **Vpv**. The motor speed should now be changing between 2500 rpm and about 3500 rpm.

Note: Your response should be similar to that shown on the right. It may vary depending component tolerances and on the type of motor used. You may need to adjust the value of **Ri** for the best response without oscillation.

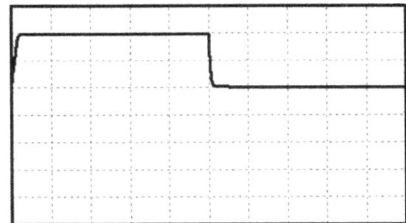

VERTICAL: 0.5V/DIV HORIZONTAL: 0.5S/DIV

Be sure that **Vpv** settles on 2.5 volts when **Vsp** = 2.5 volts.

Vpv will look "fuzzy" due to the ripple from the tachometer filter. Use the average value of **Vpv** for your data.

6. Record the following data for proportional gain of 10 and integral gain of 20 (or ?), and differential gain of 22 :

Vpv rise time: _____   Vpv fall time: _____

Vpv high settling voltage: _____

Vpv low settling voltage: _____

If your response is not what you expect, vary the PID gains (based on your experience so far) to get the best response.

Write a report on this experiment. Use a word processor and "cut and paste" to transfer all relevant information. Include the following:

1. Introduction and objectives.

2. Data and graphs for each part.

3. Analysis of data and graphs.

4. Summary and conclusion.

Application 7: Magnetic Theremin Project

Introduction

This project demonstrates an application of a linear Hall sensor IC and a photo-resistor opto-coupler.

Hall Sensor and Amplifier

The Hall effect sensor in the diagram below has a magnetic field sensitivity of about 3.125mV per Gauss. Its zero magnetic field output voltage is about one half of its supply voltage. Its output voltage increases or decreases according to the magnetic field strength and polarity as shown below on the left. The schematic diagram of the sensor and amplifier circuit is below on the right.

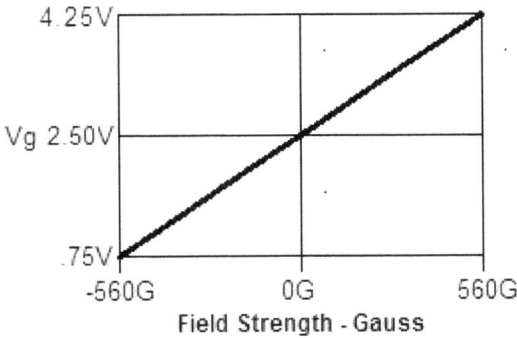

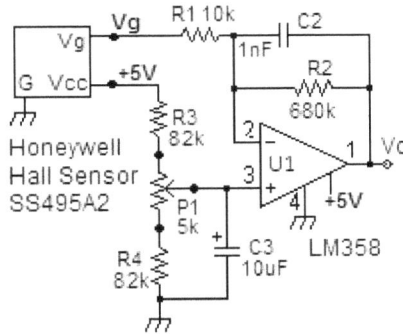

U1 amplifies the output of the sensor by 68 (680k/10k), increasing the magnetic field sensitivity to 212mV per Gauss. Its output voltage, Vo, is offset by 2.5 volts by the "balance" potentiometer, P1. A voltage divider consisting of R2, P1, and R4 is used to apply 2.5 volts to U1, pin3. The capacitor C2 in conjunction with resistor R5 provides low pass filtering.

P1 is used to set the zero Gauss offset voltage to one half of the supply voltage. In operation, this is done by adjusting it for the lowest pitch in the absence of a magnetic field. The output voltage Vo drives a voltage to current converter which is described next.

Magnetic Fields: Basic Information

The strength of a magnetic field is typically measured in Gauss or Teslas (1Tesla = 10,000 Gauss). Permanent magnets typically have magnetic field strengths ranging from about 1000 Gauss about 14,000 Gauss. The earth's

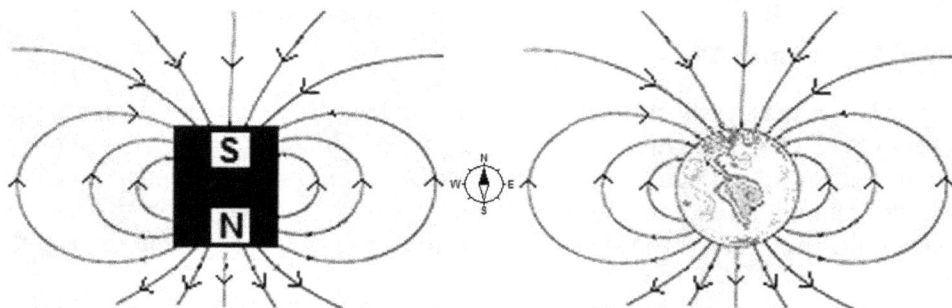

magnetic field has a typical strength of about 0.5 Gauss.

Magnetic fields may be visualized as shown in the diagrams above. The field is strongest in the direction parallel to the field lines and zero in the direction perpendicular to the field lines. Also, the field is stronger where the field lines are closer together.

The diagram above on the left shows the field lines produced by a magnet. The magnetic field is strongest at the poles and decreases with distance from the poles.

A magnetic compass needle aligns itself parallel to the field lines and its north pole points to the magnet's south pole, as shown in the middle of the diagrams above. The diagram on the right shows that the earth's magnetic field is similar to that of a magnet and that its geographic north pole is actually a south magnetic pole.

Photo-Resistor Output Opto-Couplers

A graph of typical LED current and resistance is presented on the right for the Silonex NSL-32 Opto-coupler. The diagram and drawing is presented below.

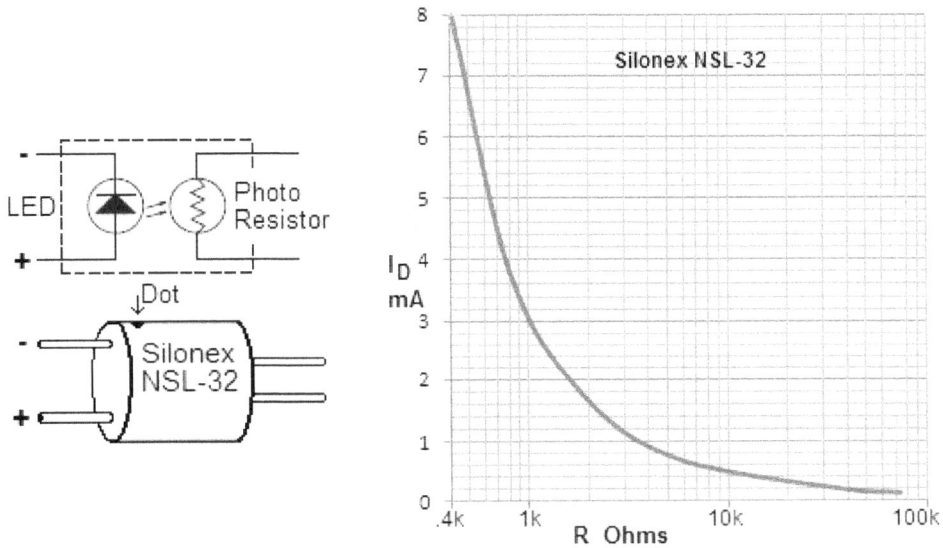

The voltage to current converter is used to vary the current through the LED of photo-resistor output opto-coupler. Increasing the LED current increases the LED brightness which in turn decreases the resistance of the photo-resistor.

Refer to the diagram and equations below. The LED current is controlled by the voltage, Vc, and by the value of R2. This circuit is a voltage controlled voltage divider. The input signal, Vin, is divided by the resistors R and R1.

$$I = \frac{Vc}{R2} \qquad Vo = \frac{R1}{R+R1} Vin$$

An increase in Vc increases the LED current. This causes the resistance of R to decrease and the output voltage, Vo, to increase.

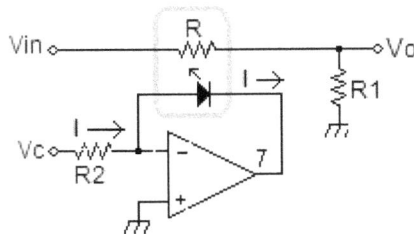

This circuit could be used in any application requiring a voltage controlled resistance, including audio and video level controls.

Voltage Controlled Oscillator Circuit

Voltage controlled oscillator circuits are also possible application for the photo-resistor output opto-coupler. The circuit below combines the voltage controlled current source circuit with a Schmitt trigger "relaxation oscillator" circuit.

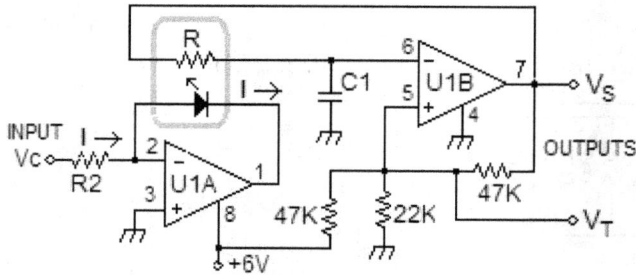

The Schmitt trigger is basically an analog comparator whose output switches between its high saturation voltage (about 5 volts in above circuit) and its low saturation voltage (close to 0 volts for an LM358).

Capacitor C1 charges when pin 7 of U1B is high and discharges when pin 7 of U1B is low. When pin 7 is high, the voltage on pin 5 is about 2.66 volts. When pin 7 is low, the voltage on pin 5 is about 1.45 volts.

The capacitor charges to 2.66 volts and discharges to 1.45 volts. The rate of charge and discharge is set by the time constant of the resistance R of the opto-coupler and the value of C1. The output of the oscillator, V_S is a 5 volt peak-to-peak square wave while the output, V_T, is approximately a triangle wave with an amplitude of about 1.2 volts p-p.

Magnetic Theremin Project Functional Block Diagram

The block diagram below shows the functional blocks of the instrument and controls associated with each block.

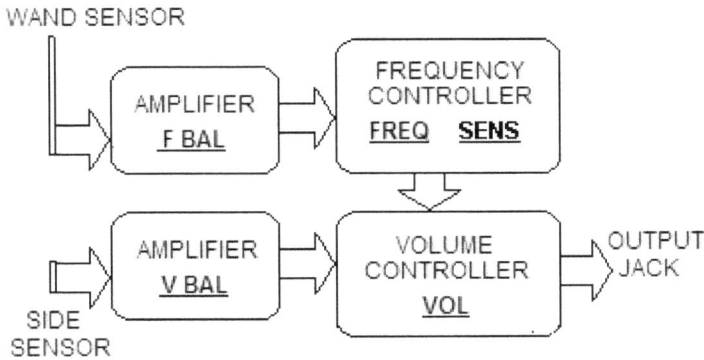

The instrument's frequency is controlled by the magnetic field detected by a Hall sensor in a hand-held "wand". The sensor's sensitivity is increased by an amplifier and it's output controls the frequency of an oscillator. The instrument's volume is controlled by the magnetic field detected by a stationary Hall sensor (side sensor).

The instrument has 5 controls whose functions are described below.

Control Functions

F BAL: This control is usually adjusted to produce the lowest pitch when the magnetic field strength sensed by the instrument's pitch wand is minimum.

FREQ: This control adjusts the range of pitch produced by the instrument.

SENS: This control adjusts the instrument's sensitivity to a magnetic field change.

V BAL: This control is adjusted to produce the highest volume when the magnetic field strength sensed by the instrument's side sensor is minimum.

VOL: This is the master volume control. It determines the maximum volume or amplitude that the instrument will produce.

Magnetic Theremin Circuit

Below is the diagram and parts list. It uses two linear Hall sensors, a voltage controlled oscillator and a voltage controlled attenuator.

Circuit Diagram

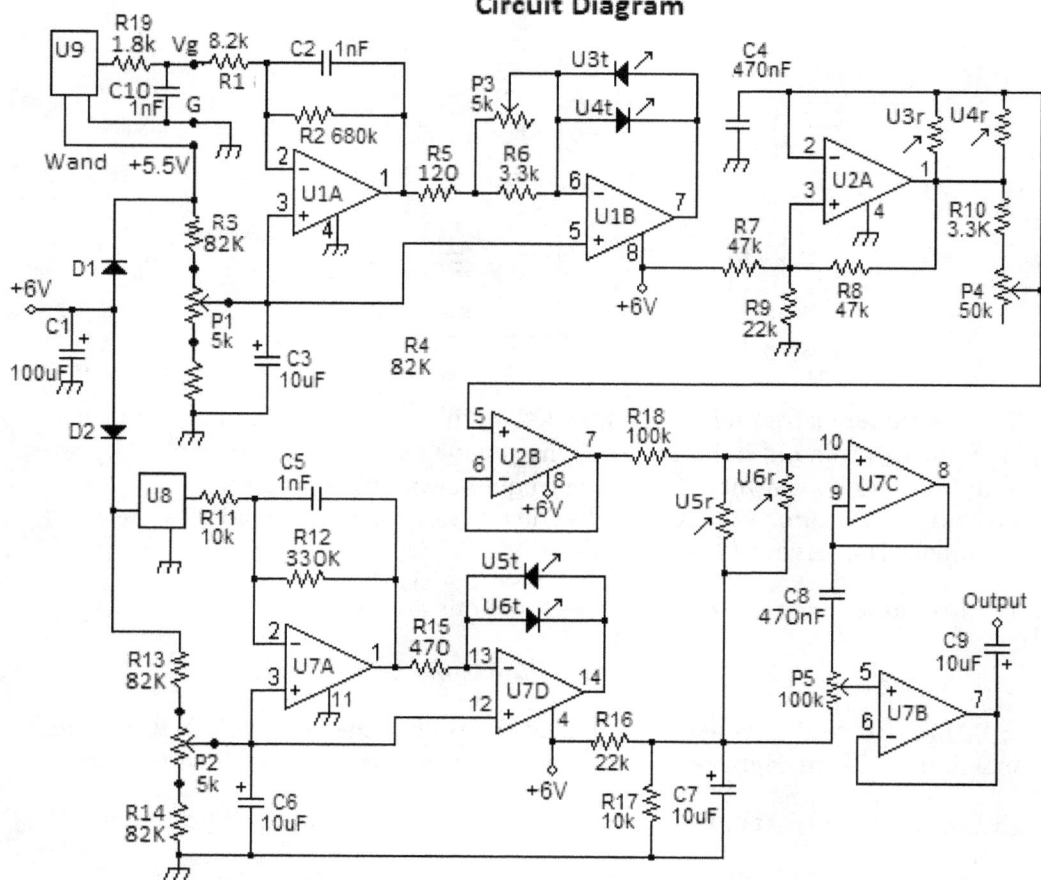

U3, U4, U5, U6: NSL-32	R11, R17: 10k
U7: LM324	R12: 330k
U8: SS495A2	R15: 470
D1, D2: 1N914A	R18: 100k
R1: 8.2k	C1: 100µF, 16VDC
R2: 680k	C2, C5: 1nF
R3, R4, R13, R14: 82k	C3, C6, C7, C9: 10µF, 16VDC
R5: 120	C4, C8: 470nF
R6, R10: 3.3k	P1, P2: 5k, Linear Pot.
R7, R8: 47k	P3: 5k, Log. Pot.
R9, R16: 22k	P4, P5: 50k, Log. Pot.

Technical Information

Pitch Wand, Hall Sensor, and Amplifier

The pitch wand uses a Hall effect sensor with a magnetic field sensitivity of about 3.125mV per Gauss. Its zero field output voltage is one half of its supply voltage. Its output voltage increases or decreases according to the magnetic field strength and polarity. The schematic diagram of the sensor and amplifier circuit is presented below.

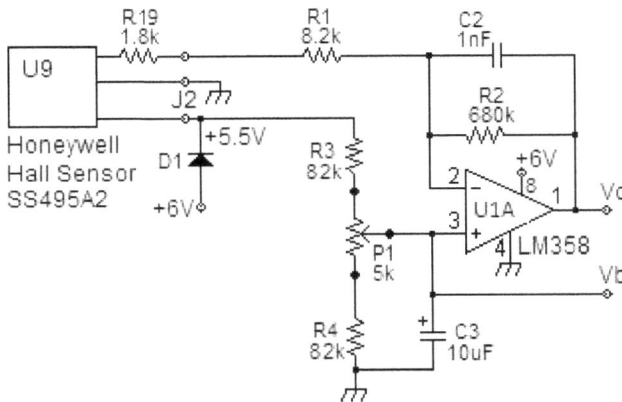

U1A amplifies the output of the sensor by 68 (680k/10k), increasing the magnetic field sensitivity to 212mV per Gauss. Its output voltage, Vo, is offset by 2.75 volts by the "balance" potentiometer, P1.

A voltage divider consisting of R2, P1, and R4 is used to apply 2.75 volts to U1, pin3. The capacitor C2 in conjunction with resistor R5 provides low pass filtering.

P1 is used to set the zero Gauss offset voltage to one half of the supply voltage. In operation, this is done by adjusting it for the lowest pitch in the absence of a magnetic field. The output voltage Vo drives a voltage to current converter which is described next. The offset voltage Vb is also applied to the voltage to current converter

Voltage to Current Converter and Oscillator

U1B in the circuit below drives current through the LEDs U3t and U4t. The current is proportional to the difference in the voltages Vo and Vb. The LEDs are the transmitter part of the photo-couplers. The corresponding receivers are the photo-resistors U3r and U4r in the circuit of U2A.

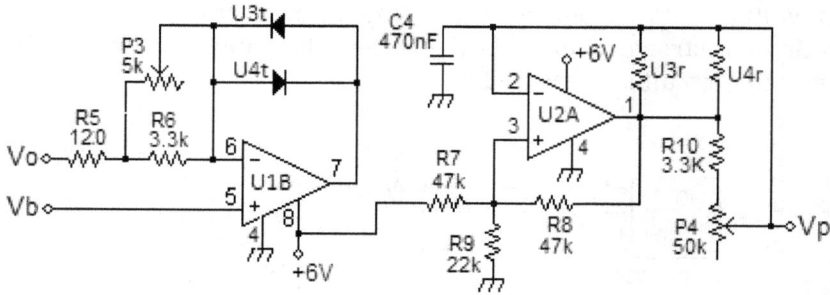

The input resistance to U1B is the net resistance of the resistors R5, R6, and P3. The net resistance can be varied from 120 ohms to 2100 ohms with P3, the sensitivity control. The voltage difference, Vo – Vb, ranges from -2 to +2 volts, depending on the magnetic field strength and polarity. Applying Ohm's law, the range of current through the net resistance is -17mA to +17mA.

When Vo – Vb is positive, the current will flow through U4t (U3t is reverse biased). When Vo – Vb is negative, the current will flow through U3t (U4t is reverse biased). Since U3r and U4r are in parallel, either U3 or U4 will cause a change in the pitch. This makes the pitch independent of magnetic polarity. Magnetic north will cause the same pitch change as magnetic south.

The sensitivity of the pitch wand to a change in magnetic field strength is calculated as follows:

1.   Sensitivity of Hall sensor is 3.125mV/Gauss. U1A amplifies this by 68.
2.   Output of U1A, Vo, changes by 212.5mV/Gauss.
3.   LED current changes by 1.8mA/Gauss when P3 is set to maximum.
4.   LED current changes by 0.1mA/Gauss when P3 is set to minimum.

The maximum change in the voltage Vo is about 2 volts. The pitch wand operates in a magnetic field range of 0 to about 10 Gauss.

The Earth's magnetic field strength is typically between 0.3 and 0.7 Gauss, depending on location. This means that the when P3 is set to maximum sensitivity moving the wand in the earth's magnetic field can produce a change in LED current on the order of 1mA, which can change the pitch by several octaves.

Oscillator Circuit

Refer to the circuit on the right. U2A is a Schmitt trigger "relaxation oscillator". The Schmitt trigger is basically an analog comparator whose output switches between its high saturation voltage (about 5 volts) and low saturation voltage (close to 0 volts).

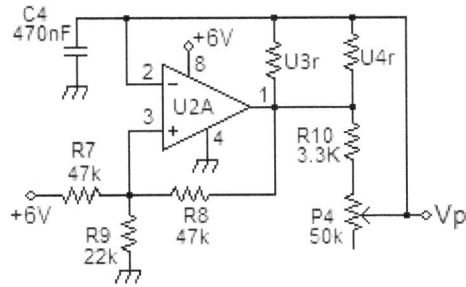

Capacitor C4 charges when pin 1 of U2A is high and discharges when pin 1 of U2A is low. When pin 1 is high, the voltage on pin 3 is about 2.66 volts. When pin 1 is low, the voltage on pin 3 is about 1.45 volts.

The capacitor charges to 2.66 volts and discharges to 1.45 volts. The rate of charge and discharge is set by the time constant of the resistance between Pin 1 and pin 2 and the value of C4. The output of the oscillator, Vp, is approximately a triangle wave with an amplitude of about 1.2 volts p-p.

The oscillator frequency is varied by the resistance of U3r and U4r. The potentiometer P4 sets the lowest frequency and effects the frequency range of the oscillator.

Volume Wand, Hall Sensor, and Amplifier

The volume wand can be any magnet. There is a Hall sensor on the side of the pc board that is used to control the output volume of the instrument. Increasing the magnetic field intensity near the Hall sensor reduces the output volume.

Amplifier, U7A, is identical to the circuit of U1A except its gain is 33. The output of U7A drives the voltage to current converter, U7D which is identical to U1B . However it has a fixed gain of about 2.1mA per volt. The total gain of U7A and U7D is about 0.22mA per Gauss.

Specifications of the NSL-32 opto-coupler indicate that its minimum photo-resistance is less than 500 ohms and its maximum (dark) photo-resistance is greater than 500k ohms.

The circuit on the right uses the opto-couplers photo-resistance in a voltage divider consisting of R18 and U5r in parallel with U6r.

Calculations show that the minimum attenuation of the voltage divider is about 3dB and maximum attenuation is about 46dB.

U7C is a buffer for the attenuator. Its output is coupled to C8 and P5. P5 is the master volume control.

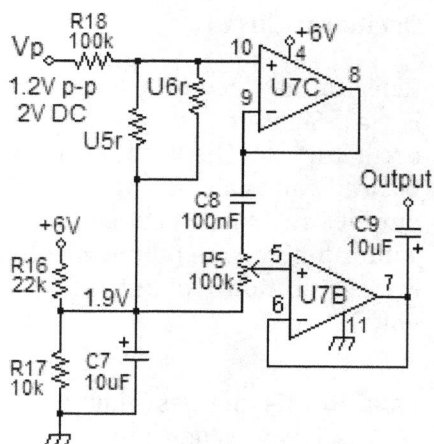

Operating Instructions

1. Turn on the power switch. Turn the master volume control (**VOL**) completely counterclockwise (-).

    Set the other controls midway (12 o'clock), slowly increase the volume. You should be able to vary the instrument's frequency with the **F BAL** control and volume with the **V BAL** control.

2. The **V BAL** control is normally set by adjusting it to produce the highest volume in the absence of a magnetic field. The volume should increase considerably at the balance point. Set **V BAL** for maximum volume and then adjust the volume (**VOL**) control for the desired zero field volume.

3. The **F BAL** control is normally set by adjusting it to produce the lowest frequency in the absence of a magnetic field. If the sensitivity (**SENS**) control is set high (+) the instrument's wand should be aligned perpendicular to the Earth's field when making this adjustment. The lowest zero field frequency can be adjusted with the frequency (**FREQ**) control.

    The settings of the controls depend on the application as well as the operator's preference and experience.

4. Summary: Adjust the **V BAL** control for the maximum volume and the **F BAL** control for the lowest frequency. Set the frequency (**FREQ**) control for the desired lowest pitch.

Move the wand near a magnet and note the resulting change in pitch. The **FREQ** control setting will also vary the pitch range produced by moving the wand.

5.   The volume can be decreased be moving a magnet or magnetic wand near the left side of the instrument (near the **V BAL** control)

6.   Experiment with the settings of the controls. The results will vary with the magnets used and their orientation. Increasing the gain will cause the wand to respond to the Earth's magnetic field.

The **F BAL** control should be adjusted for the lowest frequency when the wand is oriented perpendicular to the Earth's field. A compass may be used to determine this, or carefully moving the wand to produce the lowest pitch while adjusting the **F BAL** control will also work.

It is possible to increase the sensitivity of the instrument so that very small wand movements as well as environmental effects will cause a noticeable "waver" in the frequency. If this happens, the **SENS** control needs to be decreased (-) a little.

Magnet and Wand Information

Although any type of magnets may be used, neodymium rare earth magnets are recommended . The magnetized "volume wand" uses a 0.5 inch diameter by 0.5 inch long neodymium rare earth magnet in its tip. Larger neodymium magnets are not recommended because they can be dangerous to handle.

The earth's magnetic field or any magnet can be used to provide the magnetic field for the pitch wand, including loudspeakers and electric guitar pickups.

The "pitch wand" uses a magnetic field sensor in its tip. It is aligned so that it is most sensitive to magnetic field lines parallel to the wand. Note that the instrument's pitch can be varied by changing the angle of the wand in a magnetic field. This is the primary operating mode when using the earth's magnetic field. The variation in pitch caused by moving the wand in the field of permanent magnets depends largely on the size and strength of the magnets used and on the ingenuity of the operator.

Specifications

Frequency wand sensitivity: 0.5 Gauss to 5 Gauss
Volume sensitivity: 0 Gauss = 0dB attenuation. 5 Gauss = > 40dB attenuation
Frequency range: 20 Hertz to 8000 Hertz
Output amplitude: 1.2 Volts peak-to-peak, maximum
Output impedance: recommended load > 100Ω
Power Requirement: 6VDC, 15mA.
Batteries: 4 AAA batteries. Approximate life: 70 hours.

The circuit board and some of the parts may be available from
from ZAP Studio. Inquire at: info@zapstudio.com

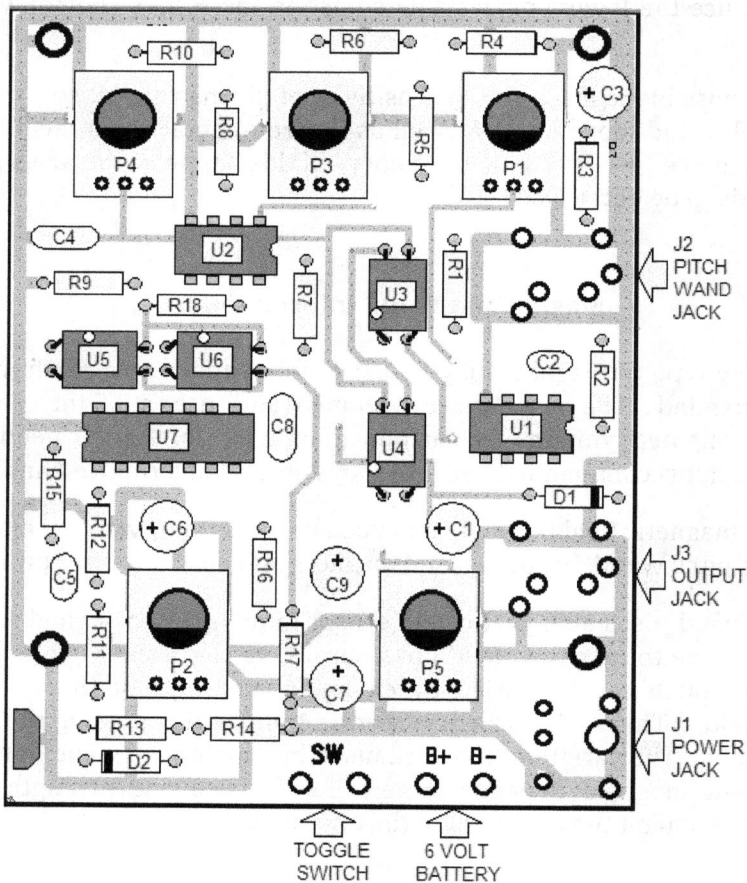

Appendix 1: Motor , Tachometer, and Oven Information

The temperature control applications in chapter 7 require a heater and temperature sensor. This may simply be a resistor with a temperature sensor near it. The control system will in effect be controlling the temperature of the resistor.

Placing the heater resistor into a small plastic box reduces the effects of air currents. The same control system could be used to control other devices such as an aquarium heater or an environmental chamber (however, the heater power circuit would need to be modified to match the heater power requirements).

The motor control applications require a dc motor with a tachometer. A motor-tachometer suitable for the motor speed control applications in chapter 7 is presented here. The dc motor is from Jameco Electronics (231781 MOTOR, DC, 3-9V, 0.25A, $1.75).

The photo-interrupter is a VISHAY TCST1202 or similar, also available from Jameco. The picture below shows the motor, tachometer, and oven assembled on a pine board for convenience. A small breadboard is included to allow for interface circuits such as a frequency to voltage converter for the tachometer and a scale amplifier for the temperature sensor.

A "photo-interrupter" is used for the tachometer. A 1.25" diameter fender washer" with a 3/16" hole diameter is used as the interrupter wheel. Four holes are drilled into it and it is attached to the motor shaft using 1/16" inside diameter tygon tubing and an LED panel mount. See picture below.

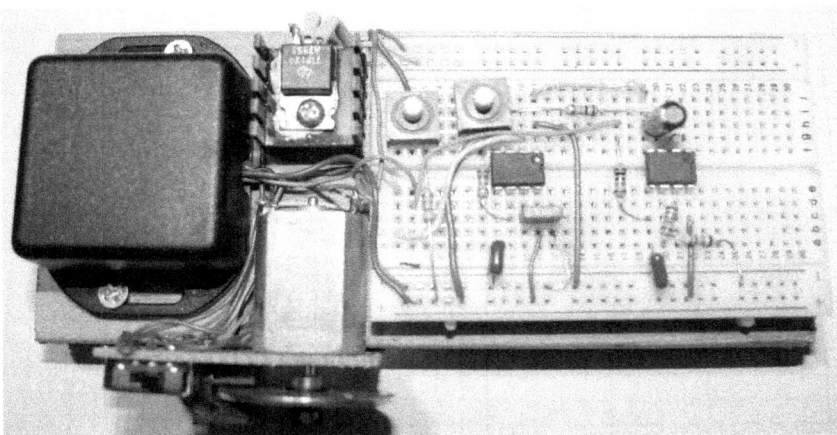

111

The holes in the tachometer wheel must be uniform and evenly spaced to avoid jitter of the tachometer waveform, **Vp**, on the oscilloscope. The duty cycle of Vp should be between 80 and 90 percent.

The diagram below shows the photo-interrupter circuit, the heater circuit with an LM35 temperature sensor, and the TIP120 darlington power transistor. The motor and the photo-interrupter circuit are mounted on a perf-board and attached to a small pine board. A slide switch connects the TIP120 to either the heater or the motor.

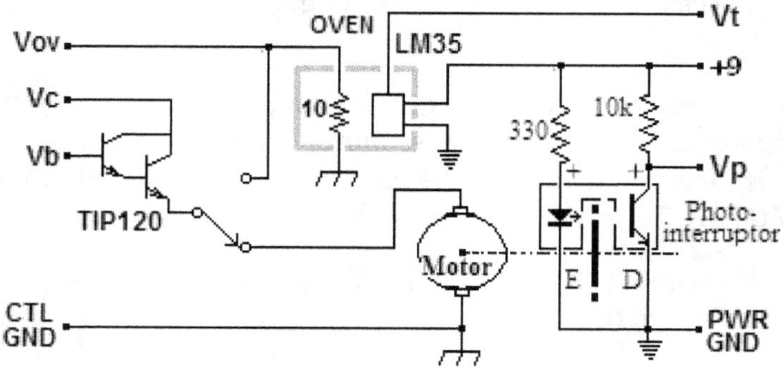

Note that the TIP120 must be mounted on a heat sink so that it is capable of dissipating at least 1 watt. The picture below shows the heater with the lid removed (Enclosure: Hammond 1551PFLBK). The LM35 sensor is shown next to the heater resistor.

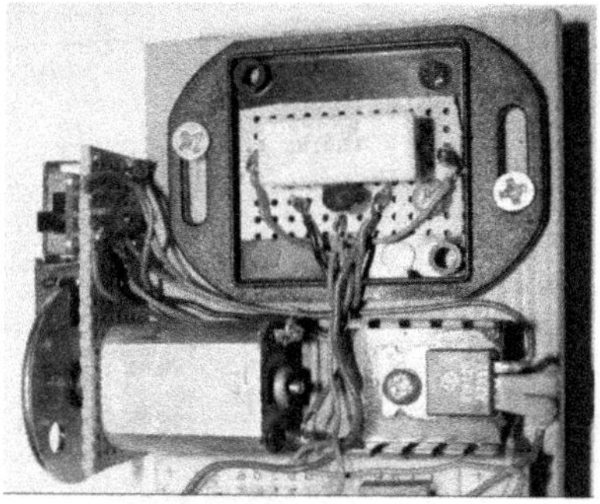

Appendix 2: PID-X1 Controller Board Information

A PID circuit board that plugs into a breadboard may be used for the proportional, proportional plus integral, and proportional plus integral plus derivative mode experiments and applications in this book. Refer to the schematic diagram below:

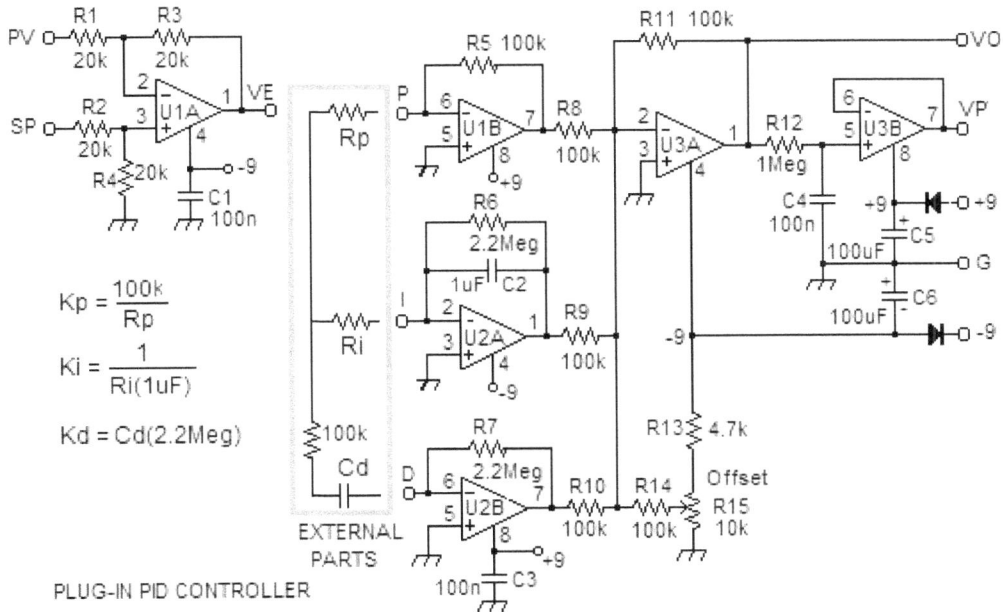

U1A is a unity gain differential amplifier. Its output is the error voltage VE, which is the difference between the set point voltage, VSP, and the process variable voltage, VPV. U1B is a proportional amplifier whose gain is controlled by Rp.

U2A is an integrator whose gain is controlled by Ri. U2B is a differentiator whose gain is controlled by Cd. U3A is a unity gain summer that sums the PID output voltages and the offset voltage. R15 controls the offset (equilibrium) voltage. The U3B circuit may be used to simulate a process with a time constant of tc = R12·C4 = 0.1 second.

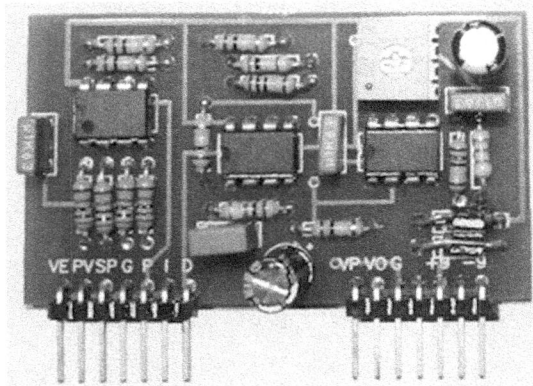

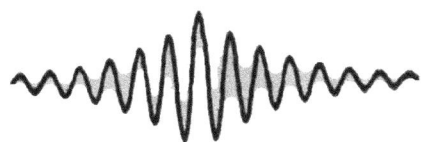

www.ingramcontent.com/pod-product-compliance
Lightning Source LLC
Chambersburg PA
CBHW051222200326
41519CB00025B/7211